Routledge Revivals

Water Management Innovations in England

As pressures on water resources have increased, problems of water quality have claimed high priority in national concern and governmental policy. In this book, first published in 1969, Lyle E. Craine describes how Great Britain enacted new governmental procedures for studying, planning, and executing water management programmes. Although the physical and social characteristics of the United States' water resources problems differ from those of England, this analysis of the British institutional arrangements for water management suggests constructive insights for managing water resources within the individual states. This title is a valuable resource for students interested in environment and sustainability issues, national water resources problems, and government policy making.

W0234852

Water Management
Innovations in England

Lyle E. Craine

RFF PRESS
RESOURCES FOR THE FUTURE

First published in 1969
by Resources for the Future, Inc.

This edition first published in 2016 by Routledge
2 Park Square, Milton Park, Abingdon, Oxon, OX14 4RN
and by Routledge
711 Third Avenue, New York, NY 10017

Routledge is an imprint of the Taylor & Francis Group, an informa business

© 1969 Resources for the Future, Inc.

Publisher's Note
The publisher has gone to great lengths to ensure the quality of this reprint but points out that some imperfections in the original copies may be apparent.

Disclaimer
The publisher has made every effort to trace copyright holders and welcomes correspondence from those they have been unable to contact.

A Library of Congress record exists under LC control number: 70075182

ISBN 13: 978-1-138-94507-4 (hbk)
ISBN 13: 978-1-315-67158-1 (ebk)
ISBN 13: 978-1-138-94508-1 (pbk)

WATER MANAGEMENT INNOVATIONS IN ENGLAND

Lyle E. Craine

RESOURCES FOR THE FUTURE, INC.
1755 Massachusetts Avenue, N.W., Washington, D.C. 20036

Distributed by
The Johns Hopkins Press, *Baltimore, Maryland 21218*

RESOURCES FOR THE FUTURE, INC.

1755 Massachusetts Avenue, N.W., Washington, D.C. 20036

Resources for the Future is a non-profit corporation for research and
education in the development, conservation, and use of natural
resources. It was established in 1952 with the co-operation of the Ford
Foundation and its activities since then have been financed by grants
from that Foundation. Part of the work of Resources for the Future
is carried out by its resident staff, part supported by grants to
universities and other non-profit organizations. Unless otherwise
stated, interpretations and conclusions in RFF publications are those
of the authors; the organization takes responsibility for the selection
of significant subjects for study, the competence of the researchers,
and their freedom of inquiry.

This book is one of RFF's studies in water resources and the quality
of the environment, directed by Allen V. Kneese. Lyle E. Craine is
professor of resource planning and conservation at The University of
Michigan. The manuscript was edited by Doris L. Morton. The charts
were drawn by Clare and Frank Ford.

RFF staff editors: Henry Jarrett, Vera W. Dodds, Nora E. Roots, Sheila M. Ekers.

Standard Book Number 8018-1027-2

PREFACE

The development and management of water resources has posed problems of an engineering, economic, and institutional character for societies and nations throughout history. Agricultural countries have had to cope with problems of both insufficient and excessive water for their agriculture. Water systems in the industrialized nations have had to be managed to provide communities with adequate water supplies, commerce with water transportation, and industries with a means of waste disposal. As these pressures on water resources have increased, problems of water quality have claimed high priority in national concern and governmental policy.

In this volume, Lyle Craine describes how one country has organized itself to undertake this job. In 1963 Great Britain enacted new governmental procedures for studying, planning, and executing water management programs. Based on local River Authorities, the approach is regional with considerable local autonomy. National co-ordination is to be maintained through supervisory authority vested in the central government.

The British pattern stands in contrast to that adopted in the United States, as do the physical and social characteristics of Britain's water resources problems. There is less climatic variation in Britain, and there is less contrast in attitudes and goals with respect to water resources than that which one experiences between the humid and temperate East and the arid West of the United States. The British river basins are geographically smaller in area and more densely settled than the great multistate basins in this country. Yet the analysis of the new British institutional arrangements for water management suggests constructive insights for managing water resources within the individual states.

After describing water management and its component processes, Professor Craine develops a set of criteria which permits him to evaluate the new British arrangements for regional water development. With occasional comparative reference to the United States, he indicates their strengths and primary weaknesses.

As our own limited water resources are facing rapidly growing demands of all sorts, studies of this type are especially useful. We have found that while good technical solutions to water management problems may be identified, they frequently are difficult to implement through the existing water resources institutions of national and local governments. Clearly, if society is to realize the potential benefits of such water management programs, we must design institutional arrangements capable of their execution. Rarely can this be done on the basis of pat formulas, given the importance of historical experience in determining public attitudes towards water resources and what societies believe are the appropriate roles for various levels of government in the management process. Rather, it is necessary to experiment with a variety of institutional arrangements.

It would prove exceedingly costly and time-consuming to experiment with all possible arrangements in any individual country. On the other hand, it is possible through comparative studies of the type represented by this report to assess the strengths and weaknesses of institutional arrangements tried in different countries. This appraisal of the new British Water Resources Act should prove illuminating to students, administrators, and responsible water managers who are concerned with putting into effect the best technical plans possible for managing water resources elsewhere in the world. The social sciences have the obligation to look beyond the problems of delineating technically desirable programs for natural resources development. They must probe the opportunities for and obstacles to the implementation of such programs. The current volume makes a significant contribution toward this end.

MICHAEL F. BREWER
Vice President
Resources for the Future, Inc.

ACKNOWLEDGMENTS

Many persons in England and the United States have aided in the conduct and presentation of this study.

For their warm hospitality while I was in England, for their sustained help in my efforts to understand the essential features of water management in England, and for their thoughtful advice and constructive criticism, I am especially grateful to: N. H. Calvert, Ministry of Housing and Local Government and first Secretary of the Water Resources Board; H. C. Darby, Professor of Geography, University College, London, and now a member of the Water Resources Board; Ian Drummond, Solicitor and Clerk of the Trent River Board and now Solicitor and Clerk of the Trent River Authority; E. A. G. Johnson, Chief Engineer, Ministry of Agriculture, Fisheries and Food; Marshall Nixon, Engineer to the Trent River Board, and now Engineer to the Trent River Authority; C. H. Spens, Chief Engineer, Ministry of Housing and Local Government; and Peter O. Wolf, Lecturer in Hydrology, Imperial College, London, and now Professor and Head of the Department of Civil Engineering, The City University, London.

For their provocative criticisms and helpful ideas in preparing and presenting this monograph, I am indebted to: Blair T. Bower, Michael F. Brewer, Edwin Haefele, Henry Jarrett, and Allen V. Kneese of Resources for the Future, Inc.; Irving K. Fox, formerly of Resources for the Future and now Professor of Regional Planning, University of Wisconsin; Robert Warren,

Associate Professor of Political Science, University of Washington; Mrs. Martin David, Department of Agricultural Economics, University of Wisconsin; and Sol Jacobson and Demitri Plessas, graduate students in Resource Planning and Conservation, The University of Michigan.

In addition, my gratitude is extended to the following people in England who contributed information and aided my study in a variety of ways: A. G. Boulton, Director of Surface Water Surveys, Ministry of Housing and Local Government; A. Key, Senior Chemical Inspector, Ministry of Housing and Local Government; Louis Klein, Chemical Consultant (formerly with the Mersey River Board), Manchester; W. F. Lester, Fisheries and Pollution Officer, Trent River Board; M. A. Lidell, River Boards Association; K. C. Knoble, Engineer, Ministry of Agriculture, Fisheries and Food; N. A. F. Rowntree, Partner, Rofe and Raffety Engineering Company, now member and Director of the Water Resources Board; D. E. Tucker, Fisheries and Pollution Officer, Bristol Avon River Board; E. B. Worthington, The Nature Conservancy; G. M. Yates, Clerk and Chief Financial Officer, Bristol Avon River Board.

For efficient stenographic and clerical assistance my thanks go to Anita Sanchez and Alice Bond of the School of Natural Resources, The University of Michigan; and to Doris Stell, Vera Ullrich, and Maria Aveleyra of Resources for the Future.

My wife, Asho, has assisted in many ways, especially by her incisive review and criticism of manuscript and her encouragement and companionship throughout.

The study would not have been undertaken and completed without the continuing encouragement and financial support of Resources for the Future. My gratitude to RFF is immeasurable and enduring.

LYLE E. CRAINE

Ann Arbor, September 1968

CONTENTS

LIST OF TABLES

LIST OF FIGURES

WATER MANAGEMENT INNOVATIONS IN ENGLAND

1

INTRODUCTION

The passage of the Water Resources Act of 1963 has given water management new meaning and impetus in Great Britain. The new statute expanded and reorganized governmental powers and activities and for the first time made possible a management approach to the water resources of England and Wales. Under the new scheme, water management is decentralized, with twenty-nine regional agencies wielding a wide range of powers for comprehensive water management. These agencies—twenty-seven River Authorities created by the 1963 Act and the Thames and Lee Conservancies, which were established earlier but which exercise essentially the same powers—are controlled by representatives of local government and are required to operate within carefully prescribed procedures supervised by two principal cabinet Ministries and a newly created Water Resources Board.

The recent British experience in organizing water management deserves careful consideration in the United States. In this country the federal government took a commanding lead in water development during the nineteenth and early twentieth centuries as a means of supporting economic growth. The emphasis was on navigation, irrigation, power, and flood control. American technology and public institutions for these purposes are now well developed. However, in the urban-industrial economies of the nation, where water management practices are becoming increasingly important, the inherited policies and

1

institutions are proving inadequate, and institutional innovations are being sought with increasing urgency.

In the United States, as in Western Europe, the term water management has come into currency during the last decade, as efforts were made to obtain more complete and rational control of water use and development. Yet the term has not been given specific meaning, and there is no common understanding of policies and institutions requisite to achieving such a goal.

This monograph is presented in the belief that the British experience may be instructive in the search for effective policies and institutions to deal with water use and development in the large urbanizing regions of the United States. However, it would be unwise to assume that British patterns are directly transferable to the American scene. No implications of "lessons to be learned from this study" are intended. Nevertheless, the British experience may be a source of fruitful ideas in developing the concept of water management in the United States, particularly within the individual states now seeking a more effective role in water resources affairs.

Scope of Study

This study describes and evaluates the policy and institutional aspects of the new British system of water management. The legal powers, the administrative structure, and the provisions for financing water resource activities are described in their historical context. The evaluation is an attempt to answer the general question: Do the legal, administrative and financing provisions provide an institutional setting favorable to water management? As a basis for the evaluation, the concept of water management is examined and elaborated. Its characteristics and the conditions necessary to its effective pursuit are identified, and institutional "criteria" are proposed.

The objective of the study is constrained in two principal ways. First, there are no effective criteria for measuring institutional adequacy. At best, institutional features or characteristics that favor water management operations can be identified. The second constraint is found in the limited investigations that have been possible. At the time the study was initiated in England, the Water Resources Act of 1963 had been passed but was not to take effect for another year, not until April 1965. Neither time nor circumstances have permitted a systematic follow-up since the law became operative. Consequently, the present study is not an appraisal of actual operating experience, but an evaluation based upon the institutional structure and procedures provided in law and in promulgated administrative policies.

Definitions

The British employ some terminology not common in American water lexicon. For the convenience of the reader, six terms widely used throughout the monograph should be explained.

1. *Conservation:* In England this word has a more specific meaning in water resource affairs than it has in the United States. It is primarily used to designate works or actions designed to store and make water available in quantities and at times suitable to the needs. Conservation works usually mean storage reservoirs, particularly those used for augmenting water supplies for municipal and industrial purposes, and works required to develop or augment aquifers.

2. *Water undertakers:* In Great Britain, water undertakers are essentially municipal water supply enterprises. Historically, these were almost wholly private enterprises; during the last three decades public authorities have also become "water undertakers." Early charters for water undertakings were obtained by Act of Parliament, thus these are referred to as statutory undertakers. Now the Minister of Housing and Local Government is authorized to "charter" new undertakings or revisions and reorganizations of certain existing ones.

3. *Abstraction:* Abstraction in England is essentially equivalent to withdrawal in the United States. Water undertakers historically were the principal abstractors of water. Today, industries, farmers, and individuals may also be major abstractors. The law defines the amount of water which can be abstracted before the abstraction is subject to regulation.

4. *Local authorities:* Usually, this term refers to the variety of local governments. Two major types are of primary significance

to this study. One is the administrative county, managed by a county council, and the other, characteristically in urban areas, is the county borough, managed by a county borough council. All jurisdictions of local government are contained in either an administrative county or a county borough.

5. *Precept:* This term is used both as a noun and as a verb. It is essentially an obligatory assessment which one governmental entity is authorized to impose upon another. The precepted unit may pay its precept obligations from whatever sources of revenues its laws permit, usually from local "rates," or taxes.

6. *Rates:* Rates are local government assessments upon property within the government's jurisdiction. Levies are stated as a percentage of value, thus "rates." Rates play the role that local property taxes do in the United States.

7. *Minister:* Several ministers in the British Cabinet may, under appropriate circumstances, have responsibilities and exercise powers relating to water management. However, under the Water Resources Act of 1963, the Minister of Housing and Local Government becomes the principal minister in water resource matters. Accordingly, in this monograph, as in the statute itself, the term "Minister" refers to the Minister of Housing and Local Government. When references are made to other ministers they are specifically designated.

2

THE NATURE OF WATER MANAGEMENT

Water management is a governmental response to the grow-ing need to maximize the productivity of specific hydrologic resources. Typically, the initial response to growing demands for water and water services has been to develop additional resources. As long as water resources were plentiful, with many relatively low-cost development opportunities, this approach generally has satisfied. As opportunities for new developments decrease, i.e., as only higher and higher cost options are available, public concern focusses upon more efficient use by existing users and the allocation of the resource among users. The need for greater efficiency in use and allocation among users calls for the application of new criteria in planning, designing, and operating water development facilities and requires various degrees of in-tervention in water uses. Water management seeks to meet these two requirements.

The purpose of this chapter is to elaborate the concept of water management in a way that will provide a basis for evaluat-ing the policy and institutional arrangements for water manage-ment in England. Although definitive institutional criteria are lacking, the present chapter will identify significant characteris-tics of water management and suggest institutional features favorable to the accomplishment of its goals.

The Objective of Water Management

Water management represents an advanced state in govern-ment's involvement in water resources. It suggests the concept of a production function in which water resources are viewed as potential inputs to a water resource management system, the outputs from which are some combination of water services. This is graphically illustrated in Figure 1.

The output of water management is a wide range of "goods and services," including water itself and the services and benefits derived from water management actions. Collectively, these are referred to as water services and are of two general kinds. First, there are those that are overtly sought, that is, the production of these services is the motivating reason for taking a specific action. Second, there are those services that are incidental to, or by-products of, providing overt services. Both the overt and incidental services may be either direct or indirect services or benefits. The production of direct, overtly sought services has the characteristics of public utility production.

Government's objective in water management is, *first*, to choose the combination of water services to be produced from specified resources, and *second*, to provide those services efficiently. Given this two-pronged objective, water management

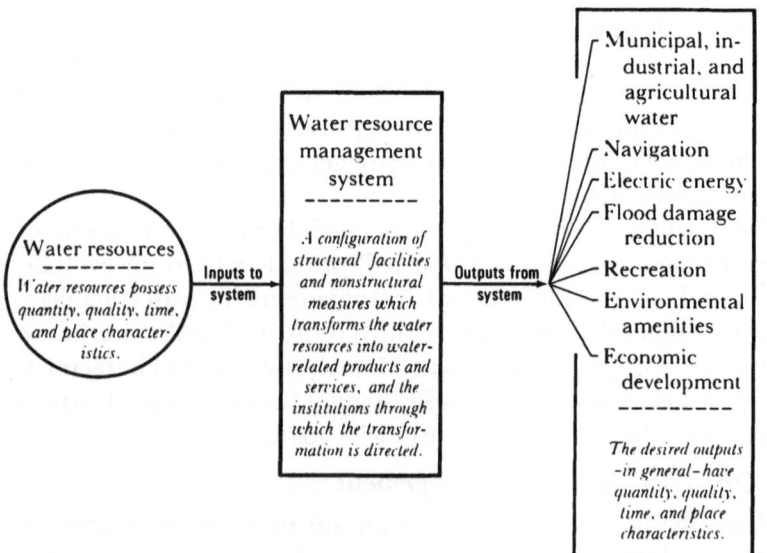

FIGURE 1. A GRAPHIC PRODUCTION FUNCTION FOR WATER MANAGEMENT. (Source: Based on suggestions of Blair Bower.)

becomes a process comprising the following activities: (1) appraising demand and supply; (2) assessing benefits and costs of alternative ways of meeting demands; (3) choosing actions which promise to meet demands most efficiently within given policy constraints; (4) implementing chosen actions; and (5) carrying out such continuing operations as are indicated.

Viewing water management as a production function is most useful and appropriate where the provision of public utility-like services is its dominant goal. Where incidental and/or indirect services are important, the same kinds of activities are essential, but they are likely to be conducted under more severe constraints.

Characteristics of Water Resources and Their Management

Many characteristics of water resources limit the effectiveness of private enterprise in providing water services and thus favor public intervention in water resources affairs. These same characteristics have the effect of setting conditions which must be met if governmental involvement is to be effective. Five such characteristics are important in giving better definition to the scope of water management and its institutional requirements.

Multiple use

Multiple-use capability of a water resource is a fundamental factor in prescribing the nature of the water management job. The concept of multiple use has two distinct aspects. In one sense, it refers to the reuse capability of any given unit of water. For example, the same gallon that generates electricity may also provide domestic and industrial water supply and/or improve water quality and recreational opportunities. Because of the multiple-use capability, any evaporation, diversion, or quality deterioration may decrease the potentiality of water services and is, therefore, of major concern in water management. Moreover the reuse capability requires water management to be concerned with the way in which withdrawal water is used as well as with the development of new sources. Allocation of water to low-priority uses, extravagance in use, and wastage in transmission are increasingly critical management factors in the efficiency of resource use. Consequently, regulations affecting allocation of water to specific use sectors and the efficiency of irrigation practices, industrial processes, and municipal distribution systems must come within the purview of water management.

Multiple use also is applied to water control structures. A reservoir may serve several purposes: to generate electricity, store flood waters, provide water recreation, facilitate navigation, supply water for domestic, industrial, or irrigation uses, or augment low seasonal flows as a water quality improvement measure. Some of these uses may be compatible, but often they are in conflict in varying degrees. Similarly, multiple-use potentials are found in ground water aquifers. Since reservoir sites and aquifers are typically limited, water management seeks the best combination of uses of each potential storage site and aquifer. It should not be assumed, however, that multiple-purpose development is always necessary in order to achieve efficiency in water resource use. Making provision for specialized uses of aquifers, impoundments, streams, or reaches of a stream may, in certain circumstances, be the preferred way of utilizing the water resource. This broader concept of multiple purpose and the choices which must be made are clearly vital characteristics of water management.

Interdependencies in the hydrologic system

Because water is a fluid that moves from higher to lower elevations, the actions of independent users or development agents are interrelated. For example, the effects of discharging a waste material into a lake or river may constrain the use that can subsequently be made of the water at other locations. Likewise, the construction of a reservoir in a drainage system changes the flow behavior and modifies water-use potentials downstream. Releases of water from impoundments alter the capacity of the stream to assimilate wastes and may reduce or increase damages associated with waste discharges below the impoundment, depending upon amount, timing, and quality of storage releases.

Given its multiple-use potentials and the physical interdependencies which characterize water resources, the hydrologic system becomes a vital factor in delineating a relevant management jurisdiction. Since the establishment of the Tennessee Valley Authority in the United States, the river basin has gained popular recognition as the appropriate physical unit for integrated water resource development. However, the river basin is not always sufficiently definitive for water management. If it is to be effective, water management must be organized for areas delineated by the actual hydrologic systems that are subjected to management actions. In some cases, a physiographic river basin may contain more than one discrete hydrologic unit, especially in arid regions or where a drainage system crosses distinct climatic regions.[1] Conversely, by constructing engineering works, man may, in effect, make one hydrologic system out of two (or more) natural drainages.

Ground water introduces some special interdependency considerations. In many areas, ground water resources greatly exceed surface water potentials. Therefore, as use pressures strain local water resources, efficiency of resource use requires a close integration of management actions relating to each supply source. Two factors underlie the need for integration. First, under pressures to augment supply, ground and surface sources should be substitutable, one for the other, in accordance with quantity and quality available, and the costs and preferences; and second, in many situations the aquifer may be hydrologically related to the surface drainage, creating thereby a single hydrologic unit. Therefore, under intense-use pressures, both hydrologic and economic interdependencies operate to urge an integrated management of surface and ground water.

Water use, regulation, and development weave a tight fabric of interdependencies, not only among different developmental purposes at a given site, but among sites on a single hydrologic unit. The conscious management of these interdependencies, based upon their social and economic consequences, is an imperative of water management. If these interdependencies are not managed, unplanned "spillover" effects from one use-site to

[1] Several of the rivers of the American Great Plains have, in effect, two identifiable hydrologic units, one in the upper arid reaches and another in the lower courses that flow through a more humid climatic area.

the next are likely to occur, which will distort the decision by each user affected. Independent water use and development agents cannot be expected to take account of the spillover costs (or benefits) which may result from their actions. Although spillovers are relevant only to the extent that their social-economic consequences are significant, water management requires that such effects be foreseen, evaluated, and incorporated into management decisions wherever their occurrence in the region is significant.

Regional disconformities

A major difficulty in organizing for the management of water resources stems from disconformities that exist among the different regional systems involved. Water management must accommodate the physical and economic imperatives of the hydrologic unit—largely those "spillover" effects stemming from interdependencies within the hydrologic system. But a hydrologic unit is, at best, only the unit of supply and seldom conforms to any one or any combination of regions within which discrete "demands" for water development services are expressed. For example, water supplies for municipal, industrial, and irrigation purposes typically are withdrawn from several different points in a hydrologic system, often by different agents, and are distributed to different clientele with different geographic jurisdiction, and may even be discharged into another hydrologic unit. Hydroelectricity may be only a part of the energy input to marketing systems extending far beyond the hydrologic region. Likewise, the "marketing" area of water-related recreation facilities is delineated by the mobility limits of the recreationists, often extending far outside the river basin and bearing little relationship to other service areas. Each service area constitutes a separate social-economic system, the "demands" from which must be articulated with the productive capacity of the hydrologic system from which it seeks an increment of service.

Similarly, jurisdictions of general government, which must be depended on for legal power, political-administrative machinery, and financial backing, seldom conform either to hydrologic or to demand regions. In some instances they may, in themselves, actually represent public service demand systems. However, their distinguishing characteristic is the overriding legal and policy constraints they may impose upon water management actions.

The disconformity of jurisdictions is the heart of the institutional problem wherever public water management is undertaken. The hydrologic system should properly reflect the production possibilities and constraints, while the demand sector should be so organized that it can express demand in consideration of production costs as well as the legal and policy constraints of the governmental system. The procedural linkages by which these components are articulated are the key to establishing institutional arrangements that will enable efficient water management.

Scale economies

Water resource development and the construction of distribution and disposal systems are capital intensive operations. Significant economies are often attainable, both in the use of the water resource and in providing water services, by introducing large scale facilities and management systems. Capturing scale economies is essential to efficiency in water management. Yet desirable scale economies often are impossible to attain under traditional institutional arrangements, both in the private and public sectors. Water management institutions, therefore, must be flexible enough to encourage, rather than discourage, large scale operations from which significant social economies may be realized.

Natural monopoly and pricing

Water resources and the production of water services are essentially natural monopolies. A market in which competitive prices are established does not exist. Consequences stemming from this characteristic are often the justification for governmental involvement in water resources.

As a governmental operation, some services are marketable and may be treated as the outputs of a public utility, but others are nonmarketable, and even an administered price cannot be levied on individual beneficiaries. Costs chargeable to these services are usually considered public costs to be defrayed from taxes. Which tax jurisdictions should pay and the amount of payment become important questions of water management policy.

Water has customarily been treated as a "free good" of nature. Charges for water are typically those necessary to cover costs of facilities required to make its exploitation possible. Under current pressures for efficiency of water use, concern is now expressed about "wastage" in municipal, industrial, and agricultural use. It is argued that if water were priced more nearly according to its value in use, losses and inefficiencies would be reduced. Such a pricing policy may be an important element in water management, particularly with regard to the public utility-like services. However, the application of such a policy will require institutional innovations.

Governmental Techniques for Effecting Water Management

In Great Britain as in the United States there is a philosophic presumption that government intervenes in water resources only to the extent necessary to assure public satisfaction with water services provided. The burden of proof is on government for any expansion of its role. Government may enter into the production function in varying degrees and through various combinations of five techniques for influencing water use and the provision of water services. These techniques are (1) water resource intelligence; (2) identification of water resource potentials and planning; (3) regulation of water use; (4) development of the water resource; and (5) organization of regional distribution (marketing) and disposal systems. The significance of these governmental techniques is discussed in the following paragraphs.

Water resource intelligence

A fundamental factor in the quality of both public and private water management decisions is the adequacy of the facts on which decisions are based. As a governmental technique to effect water decisions, water intelligence has many forms. Often one of the first activities given to government is the collection and dissemination of hydrologic data. This kind of activity tends to be the least intrusive upon individual freedoms; at the same time these data provide a common reservoir of facts and a guide to private as well as public development agencies.

Efforts to provide specific social and economic data are of more recent origin. Expansion of activities in this field, including analytic interpretations of data, represents a greater governmental involvement. At this stage, the data, the interpretations, and the publications may become much more pointed to specific problems. In such cases, intelligence activities may often directly serve some organized public "planning" effort as well as the general public and independent enterprises.

Thus the intelligence function may play a changing role along a continuum of government intervention. At one end, it could be conceived as merely providing information regarding the water resource to those who seek it, thus improving the chance of obtaining better decisions among many independent water developers. At the other end, it is conceived as a continuous network of physical, economic, and social data, including monitoring and feedback data, relevant to comprehensive water management decisions. Although the ultimate development of an information network will better serve the independent water development agents, it is of primary importance to government operations that have taken on responsibilities for water management.

Identification of resource potentials and planning

The technique of identifying resource potentials and developing plans is basically an extension of the intelligence activities outlined above. This technique may serve a variety of purposes—according to the role government chooses to play in the production function. Where government takes over as a public entrepreneur, planning should provide specifications and schedules of actions to be taken and by whom they should be taken. At the other extreme, government may choose to be involved only to the extent of inventorying resources, projecting growth in demands, suggesting alternative ways of responding to demands, and preparing reconnaissance "planning" studies. In this case, there is a presumption that the process of conducting planning studies and the dissemination of reports thereon will result in a general consensus of direction and desirable actions. Such reports, it is hoped, will provide a framework within which independent water use and development agents can proceed toward water management goals without serious conflict or losses in efficiency.[2] As a part of this notion, great emphasis has been put on the planning product, often to the neglect of the planning process as an instrument of communication and education among the interests involved.

Regulation of water use

One of the early forms of governmental intervention in water use was the enactment of codes of law establishing and enforcing water rights. Most western industrial societies by the latter part of the nineteenth century had enacted specific legislative codes designed to control the use of natural waters for waste disposal. Nineteenth century regulatory measures, never entirely satisfactory, are now grossly inadequate.

There are three principal ways of regulating the use of water resources: (1) promulgation and enforcement of standards prescribing conditions under which uses are permitted or constrained; (2) administrative allocations of water for specific uses and/or users; (3) pricing of water uses in such a way as to influence the kind, amount, and timing of use. These are not mutually exclusive but can be most effective when they supplement one another. Current measures to regulate water use deal primarily with water withdrawals and with protecting water quality.[3] Effective control over water allocations usually requires some system of permits or licenses, while regulations to protect

[2] Much of the river basin planning in the United States seems to presume this role.

[3] Licensing of fishing and boating, although generally used for other purposes, might well be extended where necessary to regulate the amount of water use for these purposes. Recent attention in the United States to flood plain zoning or other types of occupancy control, although applied to land, is also illustrative of regulation as a technique to influence water use and the provision of water services.

water quality typically rely upon a system of standards and enforcement. Comprehensive licensing of effluent discharges is a more precise control, which as yet is little used in the United States.

A management approach to water resources invites greater reliance upon use charges as a regulatory tool. To date, pricing of water services has been little used to moderate demand. However, there is a growing interest in developing schemes for more rational pricing which would attempt to relate the amount charged to the value of the water service. It is believed that water-use fees levied on domestic, industrial, and irrigation intakes, as well as charges for the use of a water body for waste disposal, would moderate demand for intake water and for waste assimilation services, would encourage greater efficiency in resource use, and would maximize the freedom of the individual user in adjusting to specific water resource circumstances.

Development of the water resource

In England and in the United States, water development has historically referred to public or private efforts to "capture" water and produce essentially a single kind of water service. Too often little attention was given to the effect of one development upon another, or to the consideration of other water services that might be produced at the same site.

Today the development technique implies a broader concept —the capacity to apply engineering and construction skills to an entire hydrologic system with the objective of maximizing the capacity of that specific resource to serve people. Water resource development, therefore, seeks to enlarge the capacity of a hydrologic unit to produce multiple services, and its goal is to maximize the net service output. This contrasts with the earlier approach to development, which merely sought to provide a specific water service for a specific population, in the most efficient way. The earlier approach leads to conflict among development agents dealing with the same hydrologic unit. The effectiveness of water management depends upon the extent to which institutional arrangements can provide development facilities that are designed and operated to avoid spillover costs and take advantage of spillover benefits associated with the interdependencies of the hydrologic system.

Three methods of providing co-ordinated development facilities are generally available: (1) direct public planning, design, construction and operation of development facilities; (2) public regulation of design, construction and operation of new projects by independent development agents; (3) contractual agreement between public agency and independent operator of existing facilities regarding operation of projects. All of these methods are presumed to be most effective if they are used to supplement one another and if they are performed by a regional water management agency that comprehends the entire hydrologic system subject to management control. The primary test of effectiveness is the extent to which significant externalities stemming from hydrologic interdependencies are considered and adjusted in planning, design, and operating decisions.

Organization of regional water distribution and disposal systems

Until quite recently, public concern about water management has principally focussed upon the water resource, that is, the quantity, quality, behavior, and availability of water in natural hydrologic units. Distribution and the "marketing" of water services, that is, the retailing aspect, and the disposal of used water have tended to be peripheral considerations.

Three facts now become significant and press for more direction over water distribution and disposal. First, water supply is a direct, overtly sought service for which there are few substitutes and for which top priority is given in allocation decisions. Second, not only the content of effluent but the manner in which it is returned to the hydrologic system may be a major factor in the use of water in other parts of the system, and therefore a major factor in the control of spillover costs. Third, the processes of withdrawing, distributing, and disposing of water effect control over a hydrologic subsystem, which is a vital part of the hydrologic system. Therefore, the design and operation of water distribution and disposal systems have a major influence upon water resources and the effectiveness with which these high priority water services are provided.

The concept that water use and waste water disposal consti-tute a subsystem of water management has been given little recognition in water resource affairs until recently. Water sup-ply and waste disposal have characteristically been handled by local independent agencies, and even within one jurisdiction they are frequently under separate management. In these cir-cumstances it is difficult to think, much less act, as though the aggregate of these operations represents a subsystem in the con-trol and use of water in its journey to the seas.

If water management is to achieve its objective, government must seek to organize these systems in such a way as to promote efficiencies in the actual distribution and disposal processes, and to contribute to efficient use of the hydrologic resource itself. Thus, government's intervention in the organization of regional distribution and disposal schemes begins to emerge as another technique for achieving water management. The objectives in the application of this technique are two: (1) to make possible scale economies in providing supply and disposal services, and (2) to better integrate the operation of these facilities with de-velopment of the resource unit. Considering water supply and waste water disposal as an integrated system is particularly im-portant in densely populated regions. Many opportunities for economies of scale can be obtained merely by consolidating or co-ordinating the conventional but independent water supply and sanitary services that are provided by the adjacent local governments of an urban region. More importantly, larger oper-ations open the possibilities of utilizing technologies that are feasible only on a large scale.

Achieving co-ordinated distribution and disposal systems de-pends upon three kinds of authorizations: (1) authority to de-sign, construct, and operate a regional system; (2) authority to acquire existing systems and to integrate their operations as indicated; and (3) authority to direct operations and to require interconnections and sharing of common facilities among inde-pendent systems.

Five techniques of governmental involvement in water re-sources have been described: providing water resource intelli-gence, identifying resource potentials and planning, regulating

water use, developing water resources, and organizing regional
water distribution and disposal systems. These techniques have
a general progressional relationship in extent of responsibility
and degree of intervention. This relationship is graphically
shown in Figure 2. There tends to be a progression in complexity and sophistication of activities supporting each technique.
For example, intelligence may progress from simple collection
of hydrologic data, to the development and analysis of social-
economic information relevant to water use and development,
and as it moves into interpretation of data and trends, it may
merge with "identification and planning."

The early stages depicted in Figure 2 are largely devoted to
creating an information and/or a policy environment that will
influence private and public water users and development
agents toward governmental goals. In the more advanced stages
(stages 4 and 5), the "intelligence" and "identification and planning" techniques may in many situations be more important as
inputs to "regulation," "resource development," and "regional
distribution and disposal," and may actually become subsumed
under a planning activity that provides the basis for integrating
the latter three techniques.

Thus, government may enter into the production function in
varying degrees by utilizing these five techniques in a variety of
combinations. The extent and kind of involvement will depend
upon specific circumstances of time and place. At one extreme,
the government may choose to perform much of the production
task directly (stage 4) and even enter into regional distribution
and disposal schemes (stage 5). At the other extreme, government may choose merely to provide a policy environment—
"intelligence," "identification and planning," "regulations"—
to influence private entrepreneurs, local governments and other
agencies to operate independently within this policy framework.
In practice, government acts through any combination of techniques on a continuum between these two extremes.

Water management is an advanced stage of governmental involvement and is generally characterized by stages 4 and 5. The
public utility approach tends to characterize the water manage-

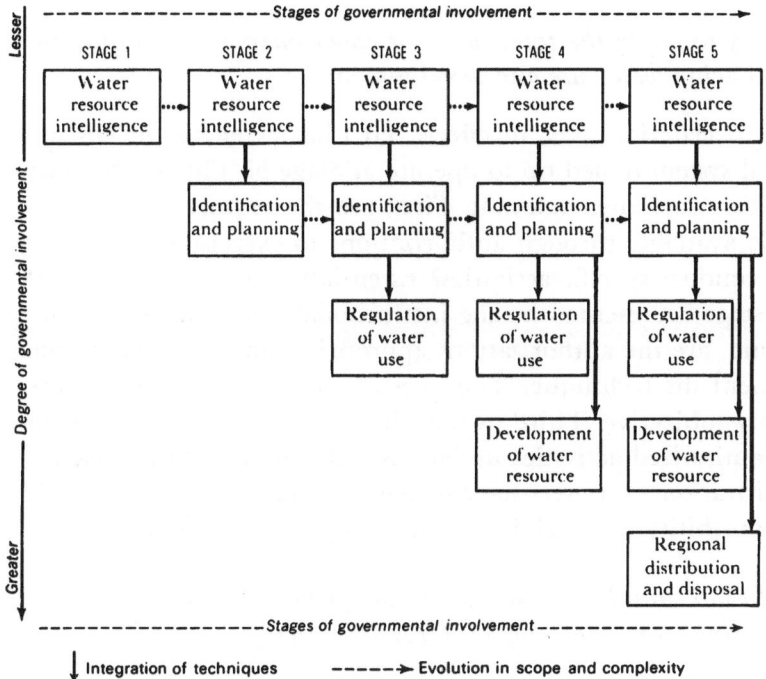

FIGURE 2. GOVERNMENTAL TECHNIQUES FOR INFLUENCING WATER USE AND DEVELOPMENT.

ment stages. Here, the classical production function pretty well describes the dominant activities; the authorized activities that support each technique are administratively integrated and become essentially a sequence of production activities. At this stage they may be directed toward a chosen set of common goals.

Criteria for Evaluating Water Management Institutions

The above descriptive analysis of the nature of water management provides a basis for suggesting five essential characteristics of institutional arrangements. These are not rigorous criteria; however, for convenience they will be referred to as institutional criteria. The five criteria will be elaborated in the remainder of this chapter and will serve in Chapter 6 as a basis for evaluating institutions for water management in England.

Ability to apply the total range of governmental techniques for influencing water use and development

This criterion asks, in effect: To what extent is the institutional system tooled up to operate at Stage 5? Three subsidiary questions are then relevant. First, are the required techniques made available through authorizations to exercise legal powers and conduct specific activities? Essentially, this is a question of the stage represented by the institutional system under review. Second, are the authorizations appropriate and adequate to implement the technique? The question of adequacy is, of course, highly subjective. Third, is the administrative responsibility for the authorized activities under unified command? Effective administration of water management can hardly be expected if responsibility for applying the techniques is dispersed.

Ability to consider and adjust (or adapt to) externalities stemming from hydrologic interdependencies

This criterion relates to the popular concept of unified management of the river basin. The essential ingredient in the concept is the ability to adjust the externalities that might otherwise give rise to inefficiencies under fragmented responsibilities.

There are two aspects to fulfilling this criterion. First, potential externalities may be institutionally internalized. That is, water management activities of an entire hydrologic system may be put under a single organization, and each organizational decision will thereby be required to confront externality questions and seek to adjust them. However, institutional internalization can seldom be complete except where political forces permit the establishment of an Authority with absolute powers over all water activities in the basin. Thus a second means of adjustment is often more acceptable and for some purposes equally successful. This involves legal and administrative requirements for the exchange of payments among water use and development agencies in accordance with spillover damages incurred or spillover benefits received.

Flexibility to adapt water management actions to different circumstances of time and place with protection against arbitrary and capricious actions

Three considerations enter into an evaluation of flexibility: (1) the degree of administrative discretion provided in the law; (2) the extent that specific decisions are constrained by administrative goals, policies, and standards; and (3) the extent to which decisions can be reassessed and adjusted over time. In judging flexibility, consideration must be given to the importance of maintaining a balance between flexibility and stability. Water management seeks to redress a past emphasis on stability. Nevertheless, a satisfactory management system should provide adequate security to independent entrepreneurs and development agencies without creating an operational straitjacket that restricts any shift in the patterns of resource use.

Ability to express and consider the range of values relevant to a water management decision

This criterion essentially is concerned with the decision making problem associated with regional disconformities among the hydrologic, demand, and governmental systems. The question is, "How can the various kinds of values associated with each system find expression in the water management decision?" Six institutional features may contribute to meeting this criterion:

1. The jurisdiction of the regional water management agency may be extended in such a way as to incorporate within its area many of the relevant hydrologic, demand, and governmental systems. In complex situations, it is not likely that any one area can be devised to include all relevant systems without incurring other overriding difficulties.

2. Pricing and inter-system transfer payments may provide linkages between systems to articulate values amenable to monetary expression.

3. Formal interagency review procedures may provide linkages through which non-monetary values find expression in the decision process.

4. The nature of representation provided on governing boards is a vital aspect in assuring that all relevant interests participate in decisions.

5. Rules by which decisions are made within agencies are also an important aspect in determining whether values are properly considered in decisions.

6. In many instances, requirements for consultations in formulating proposals and provisions for appeals and hearings may be the last resort for an individual or group seeking to be sure that its point of view and value is given consideration.

Ability to finance water management consistent with its objective of efficiency

The financial ability of an institutional system depends on: (1) whether there are any unusual obstacles to raising capital and operating funds; and (2) the extent to which disconformities in the incidence of costs and the incidence of benefits encourage inefficiencies.

The extent to which water management is recognized and built into government as a continuing function

This criterion suggests a searching skepticism of ad hoc arrangements, particularly when such arrangements do not lead to permanent legal and administrative measures that employ the five basic governmental techniques for directing water use and development. Other evidence of building water management into government may be found in procedures for relating water management decisions to other relevant governmental policies and operations.

3

ENGLAND'S WATER RESOURCES AND INSTITUTIONS

A new comprehensive water policy designed to foster water management became effective for England and Wales on April 1, 1965. The Water Resources Act of 1963,[1] on which this innovation is based, represents twenty years' effort to establish a comprehensive policy and institutional structure.

The 1963 Act applies only to England and Wales, presumably because they are the only hydrologically related political units of Great Britain. Scotland, being separated from England by the Cheviot Hills, is virtually self-contained with regard to fresh water resources. Before the 1963 Act came into full effect, an order[2] transferred the executive powers regarding water management in Wales to a Secretary of State for Wales, a new cabinet post in the central government in London. This added administrative complication is not of great significance to the central concern of this study. Therefore, in order not to blur the essential elements of the policy and institutional provisions of the Water Resources Act, the presentation that follows will refer to England, recognizing that, except for the assignment to the Secretary of State of certain administration actions, the same pattern of operations applies to Wales.

This chapter will first describe briefly the water resources situation in England and Wales and then outline the major policy and institutional arrangements provided by the 1963 Act.

[1] *Water Resources Act of 1963* (HMSO, 1963), Ch. 38.
[2] Statutory Instruments, No. 319, 1965 (Feb. 26, 1965), (HMSO, 1965).

The Water Resources of England and Wales[3]

England and Wales have a combined area of 58,343 square miles, which is comparable to the area of Georgia, Illinois, or Michigan in the United States. In terms of water resources, England and Wales may be divided in two by a line running from Weymouth on the southwest Channel Coast approximately northeast to Hull on the estuary of the Humber. Northwest of this line the underlying bedrock is mainly impervious, while in the southeast, water-bearing strata are common. Rainfall decreases from northwest to southeast, from average annual rainfalls of over 50 inches in parts of the Pennines and the Welsh mountains to the low 20's in the southeast, with local areas averaging less than 20 inches.

The physiographic, geologic, climatic, and the demographic-economic variations mark out four water resources zones, each with distinctive water use problems and development needs. The four zones are: Wales and West Midlands; Pennines; South West; and South East. (See Figure 3.)

The Wales and West Midlands zone is divided into two sub-zones by the north-south Cambrian Mountains in Wales. The western slopes drain into the Irish Sea, St. George's Channel, and Bristol Channel, while the eastern slopes of the Welsh highlands are hydrologically linked with England by the Rivers Dee, Severn, and Wye. Most of the zone is underlain with impervious rock, which, together with abundant rainfall, gives great importance to the surface water resources of the region. The Welsh Advisory Water Committee reported in 1961 that with proper development, an exportable surplus of 450 million gallons per day might be made available in the year 1990.[4] The West Midlands area of England is especially favored since it can receive through the natural river systems most of the exportable surplus of Wales.

[3] Description of the character of the water resources of England and Wales, unless otherwise indicated, is based upon Norman J. Pugh, "Water Supply," in *Conservation of Water Resources in the United Kingdom* (Proceedings of Symposium, Oct. 30–Nov. 1, 1962), The Institution of Civil Engineers, Great George St., London, S.W.1.

[4] Pugh, *op. cit.*, p. 11.

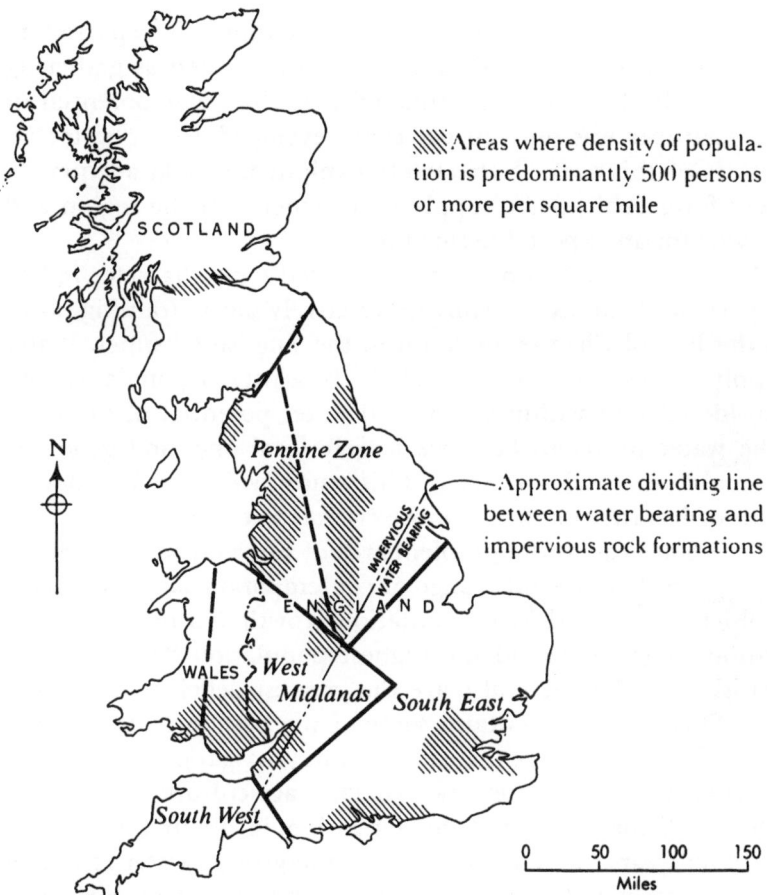

FIGURE 3. WATER RESOURCE ZONES. (Source: Adapted from *Conservation of Water Resources in the United Kingdom*, The Institution of Civil Engineers, London, 1963.)

The Pennine zone is also subdivided into east and west sub-areas, the Pennine Mountains forming the partition. Both are mainly underlain by impervious rocks, and underground water resources are in general not of importance. The Northern Pennines record some of the highest rainfall in England, and much of this area has been set aside as water gathering grounds by the many statutory water authorities serving the population centers within the service area. For example, the Peak District

National Park in the heart of the Pennines is reported to have two-thirds of its 542 square miles dedicated as gathering ground.[5] In 1958, it was estimated that the water potential of the Pennines was only 20 per cent developed, so the water resources of this area, if effectively exploited, should adequately meet foreseeable water supply requirements in the region and provide for an exportable surplus.[6]

In general, the Southwest is an isolated zone, characterized by impervious bedrock. Streams are relatively short, draining north to the Bristol Channel or south to the English Channel. Water supply requirements of the relatively scattered population are considered well within the overall water potential of the zone. The water problems here are largely local ones and generally amenable to local solutions. This study will therefore devote most of its attention to the more critical water regions.

The Southeast is the problem area of England with respect to water as well as a whole range of contemporary social-economic problems associated with population growth and urban concentration.[7] Here is found the highest population density in the British Isles, but here also are the greatest deficiencies in rainfall. This zone also contains some of the best agricultural land of England, and the expansion of spray irrigation has become a major factor in water use. In fact, agriculture around the Thames Estuary profits from irrigation in about nine out of every ten years.[8] It is reported that underground water resources in the area have largely been tapped and in certain areas, notably in the chalk formation around London, have already been overpumped.[9] In short, this area contains the greatest concentration of water demand, possesses the lowest average rainfall, is already using its ground water at a faster rate than it is replenished, and is geographically isolated by some distance from areas of exploitable surplus water resources.

During the period 1955–59, a sub-committee of the govern-

[5] Interview, John Foster, Planning Officer, Peak District National Park.
[6] Pugh, *op. cit.*, p. 11.
[7] Ministry of Housing and Local Government, *The Southeast Study, 1961–1981* (HMSO, 1964).
[8] Office of the Minister of Science, *Irrigation in Great Britain, A Report of the Natural Resources (Technical) Committee* (HMSO, 1962), p. 1.
[9] Pugh, *op. cit.*, p. 11.

ment's Central Advisory Water Committee undertook an over-all assessment of the water resources situation in England and Wales. This committee compared estimated developed water supplies in 1965 with the estimated mean yield[10] of eighteen hydrometric areas of England and Wales.[11] It was reported that more than 10 per cent development of mean yield was expected in only seven of the eighteen areas by 1965. Four of these areas are within the Pennine zone, and, taken together, expected about 19 per cent development by 1965. The other three areas are the heart of the South East zone, and, taken together, expected 31.8 per cent development by 1965.[12] Again, it should be cautioned that these figures are not quantitative absolutes since, as indicated in the sub-committee report, "mean yield" is something in excess of potential development capacity.

These two assessments of water availability suggest, as is so often the case, that the primary water supply problem is one of distribution rather than an absolute shortage, and that the distribution deficiencies are geographic as well as seasonal.

The question of water availability is further aggravated by serious water pollution, particularly in the populous and industrial centers of the Midlands and the Southeast region. Pollution has plagued England since the industrial revolution. Not until the enactment of new regulatory measures in 1961 did the law give promise of effective abatement and possible reduction of pollution in the face of increased waste generation. Perhaps because of the new pollution control law, the problems of water quality received relatively little attention in the considerations leading up to the Water Resources Act of 1963.

Concern about water quantity seemed to dominate considerations of water resource problems in 1962. Maldistribution was seen as the principal obstacle to having available enlarged sources of supplies, and emphasis was given to the need for a policy and institutional structure that would facilitate two types

[10] Mean yield is defined as "the flow of water from the area (rainfall minus calculated evapo-transpiration losses) averaged over a long period of time." No specific definition is given of development, but from context it seems to reflect estimates of amounts abstracted.

[11] Central Advisory Water Committee, Sub-Committee on the Growing Demand for Water, *First Report* (HMSO, 1959).

[12] *Ibid.*, p. 27, App. II, Fig. 7.

of development approaches. The first approach required authority to plan and carry out schemes for unified water management
within appropriate natural drainage areas. Through such measures, construction and operation of structures to regulate run-
off for multiple purposes, integrated with ground water recharge projects and the exercise of regulatory authority, could
augment total productive capacity of each drainage. The second
type of development approach required authority to plan and
control inter-basin transfers of water. Although a national
"water grid," comparable to an electric power grid, was not considered economically sound in the foreseeable future, it was
recognized that in many situations the transfer of surplus water
to reasonably adjacent areas of need would be required.[13] A few
major instances of such transfers are already in effect. Birming-
ham obtains its supply from reservoirs on the headwaters of the
Wye, 74 miles from the city; Liverpool brings water from the
Severn River basin 68 miles away; and Manchester reaches up
into the famous Lake District, over 100 miles to the north.
Additional supply for Liverpool has recently been developed in
the River Dee in northeastern Wales where it is stored and released to be passed downstream to a point of abstraction only
15 miles from a connection to Liverpool's existing supply system. Major future development schemes will undoubtedly feature diversions from the Severn Valley to the Thames Valley
whereby the exportable surplus in Wales and West Midlands
zone can be transferred to the South East zone.[14]

Unified management of surface and ground water by natural
drainage areas, co-ordinated with interregional transfer schemes,
represent the challenge to which the 1963 Water Resources Act
was directed.

Major Policy and Institutional Features

The passage of the Water Resources Act of 1963 brought
together two heretofore separate thrusts in water policy in Eng-

[13] Central Advisory Water Committee, Sub-Committee on the Growing
Demand for Water (the Proudman Committee), *Final Report* (HMSO,
1962).
[14] Pugh, *op. cit.*, p. 12.

land. One was primarily motivated by the question of adequacy of water supply for municipal, industrial, and agricultural purposes. The other policy thrust stemmed from concern about river conditions—the quality and behavior of water in the drainage basins.

Postwar policy in Great Britain has been primarily directed to achieving a greater government involvement in water supply and better co-ordination of water supply considerations with the earlier established river development responsibilities of government. In 1943, the Central Advisory Water Committee of that time (the Milne Committee)[15] expressed, for the first time in an official report, the need for co-ordination of river use and development activities in meeting the problems of pollution and assuring adequate water supplies for an expanding urban society. Yet, within five years Parliament enacted two separate and independent laws concerning water resources. The Water Act of 1945 was basically concerned with the water supply industry and placed most of the governmental powers in the Ministry of Health (later transferred to Ministry of Housing and Local Government). The River Boards Act of 1948 consolidated the then existing functions of land drainage, flood control, fisheries, and pollution control (and in some cases navigation) in thirty-two River Boards[16] operating under the joint supervision of the Minister of Agriculture, Fisheries and Food, and the Minister of Housing and Local Government. It remained for the Water Resources Act of 1963 to integrate the purposes, approaches, and methods that had become associated with these two earlier laws under a concept of water management.

The 1963 Act establishes a highly decentralized policy and administrative system for comprehensive water management, one with explicitly prescribed procedural linkages to central government, through which it is expected that national and in-

[15] Central Advisory Water Committee (Milne Committee), *Final Report* (HMSO, 1943).

[16] The Thames River Basin and Lee River Basin were earlier organized as "conservancies" and are excepted from such general legislation relating to river management as the Water Resources Act of 1963.

terregional interests will be expressed and broad policy supervision be exercised. Three major agencies of central government and a series of twenty-seven local River Authorities,
encompassing all areas in England and Wales, constitute the
principal elements of the administrative system. The Thames
and Lee Conservancies, operating under earlier legislation, are
for most purposes comparable to River Authorities and are considered as such, thus making twenty-nine regional water management agencies. (See Figure 4.)

In central government, the Ministry of Housing and Local
Government and the Ministry of Agriculture, Fisheries and
Food have long been the dominant cabinet agencies concerned
with water. Historically, the Agriculture Ministry has been responsible for land drainage, flood control, and fisheries. The
Ministry of Housing and Local Government, on the other hand,
has come into water resource affairs from the perspective of
municipal water supply needs and sewage disposal problems.
Under the new structure, each continues to play a major role
in its respective areas of responsibility, and each shares general
responsibility to promote the conservation and proper use of
water resources and the provision of water supplies in England
and Wales.

A third agency of major significance is a newly established
Water Resources Board, placed in the Ministry of Housing and
Local Government for administrative purposes. The law gives
the Board wide ranging water policy and planning responsibility and makes of the Board the chief instrument for co-ordinating local water management schemes with national and interregional needs, including inter-basin transfers of water.

River Authorities created by the Water Resources Act of
1963 are central to the governmental system of water management in England. They are basically regional agencies of water
management, their managing boards being composed of appointees of local governments and of relevant ministries in such
proportions as always to assure local government appointees a
bare majority, but no more. Each of these twenty-seven local
agencies possesses the authority to carry out basic governmental

FIGURE 4. AREAS OF RIVER AUTHORITIES IN ENGLAND AND WALES. (Source: Water Resources Board.)

powers relating to water management.[17] It is in the River Authorities where the grass roots issues of water management are faced.

Background of the Water Resources Act of 1963

Some of the influences that shaped the 1963 Act stem from the progressive involvement of government in water resources

[17] These powers are discussed in detail in Ch. 4 of this study.

throughout the past century or more; others derive more directly from various attempts to achieve co-ordination and unification of water resource efforts in the postwar period.

Prior to World War II, with few exceptions, a policy of laissez-faire appears to have dominated water use policy in England. Until the 1940's government had intervened significantly in only three aspects of water resources: navigation, land drainage, and public water supply and sewage disposal.

First, there was an early expression of national interest in the protection of navigation. This led the Crown to act as conservator of navigable waters. In *The Law of Rivers and Water Courses*, A. S. Wisdom notes:

> The King as Lord High Admiral is conservator of all ports, havens, rivers, creeks, and arms of the sea and protector of the navigation thereof, and the Crown under the royal prerogative exercised a degree of jurisdiction over the "royal rivers." In the course of time, the conservancy in many rivers has devolved upon various public bodies, whether they be commissioners, conservators, municipal corporations, navigation authorities and companies, port or harbour boards or other bodies corporate.[18]

A variety of river conservancies and specific harbor and navigation authorities today carry forward and exercise the Royal prerogative relating to navigation.

With respect to land drainage, a second aspect of water resource management characterized by early governmental intervention, Wisdom comments as follows:

> It is part of the prerogative and duty of the Crown to preserve the realm from the inroads of the sea and to protect the land from the inundation of the water for the benefit, not of an individual, but of the whole Commonwealth. Commissioners of sewers were established for that purpose in 1427, and later in more permanent form under the Bill of Sewers of 1531.[19]

Later, drainage authorities were authorized by the Land Drain-

---

[18] A. S. Wisdom, *The Law of Rivers and Water Courses* (London: Shaw and Sons, Ltd., 1962), p. 64.

[19] *Ibid.*, p. 29.

age Act of 1861. In addition, special acts created drainage authorities for particular districts. These locally controlled authorities characterized public action in drainage matters until the Land Drainage Act of 1930. The 1930 Act, as amended, now provides a complex system of technical and financial assistance from central government for agricultural land drainage and main river flood control benefits.

By the middle of the nineteenth century, government became involved in a third aspect of water through its concern for water supply[20] and sewage disposal in the face of mounting plagues and epidemics. The industrial revolution, the resulting concentration of population, and the increased use of flush toilets had conspired to make the pollution of streams a very serious public health problem, particularly in densely populated areas such as Lancashire, Yorkshire, the Midlands, and London. Dr. William Budd describes conditions of the Thames during the years 1858 to 1859:

> For the first time in the history of man, the sewage of nearly 3,000,000 people had been brought to seethe and ferment under a burning sun in one vast open cloaca lying in their midst. The result we all know. Stench so foul, we may well believe, had never before ascended to pollute this lower air. . . . For many weeks the atmosphere of Parliamentary Committee rooms was only rendered tolerable by the suspension before every window of blinds saturated with chloride of lime, and by the lavish use of this and other disinfectants.[21]

The epidemics in the nineteenth century brought the public to associate diseases and the defective state of water supplies. This led to the Waterworks Clauses Acts of 1847 and 1863, and the Rivers (Prevention of Pollution) Act of 1876, some of the first significant sanitation legislation.

The Waterworks Clauses Acts prescribed conditions under which the private water supply entrepreneurs—known as water

[20] The general evolution of governmental involvement in water supply is outlined in Central Advisory Water Committee, *Report of Sub-Committee on Water Charges* (HMSO, 1963).

[21] Quoted by Louis Klein in *Aspects of River Pollution* (New York: Academic Press, Inc., 1957), p. 2.

undertakers—were to conduct their business. Later, the Public Health Act of 1875 gave general powers to urban and rural sanitary authorities to supply water, and this Act incorporated many of the powers contained in the Waterworks Clauses Acts. The provisions relating to water supply in the 1875 Act were re-enacted with some modifications in the Public Health Act of 1936, which continued as the statutory power for many of the modern water supply undertakings by local authorities until the passage of the Water Act of 1945.

The Rivers (Prevention of Pollution) Act of 1876 (as amended) was the basic pollution control law for seventy-five years. The 1876 law was based upon a philosophy of absolute prohibition. It declared illegal the discharge of any polluting substance into the natural water body and provided legal procedures for the prosecution of violators. In spite of its "absolutist" policy, the Act never successfully combated the pressures of society. H. D. Turing in 1950 spoke of it as a "dead letter,"[22] pointing out the factors contributing to its failure. Although pollution abatement control was strengthened through fisheries legislation in the 1920's[23] and the organization of the River Boards in 1948, no fundamental changes were made in pollution control laws until 1951. The Rivers (Prevention of Pollution) Acts of 1951 and 1961 replace the 1876 law with more explicit policy declaration and comprehensive authority for administrative licensing of all effluent discharges. Before World War II, governmental intervention in water supply was timid, and pollution abatement laws were without teeth.

By mid-twentieth century, however, all types of water problems had become more acute, and more effective government actions were demanded. In general, the concern about the adequacy of water supply for an expanding and urbanizing society became the overriding motivation for expanding and unifying governmental efforts during the last two decades. Changes in governmental policies during this period were in response to

[22] H. D. Turing, *River Pollution,* The Buckland Lectures for 1950 (London: Edward Arnold and Co.).
[23] *The Salmon and Freshwater Fisheries Act of 1923* (HMSO, London), Secs. 13 and 14, Geo. 5, Ch. 16.

four factors which seem to have dominated the concern about water at mid-century. They were:

1. The question of the actual adequacy of water resources to meet the rapidly escalating "demands" for domestic and industrial water supply in the growth centers of the nation.[24]

2. The mounting severity of water pollution, as it affected available domestic water and water quality for recreation and scenic amenities to which higher values are attached by postwar society.

3. The question as to whether the traditionally independent statutory water undertakers were an adequate instrumentality to meet the mounting "demands" for efficient development and distribution of water for domestic and municipal uses. Moreover, their independence from responsibility for disposal of waste did not naturally ally them in the fight to control pollution.

4. The growing "demands" for more water and the expanding waste loads aggravated physical interdependencies inherent in water use and made compelling the integration of action for water supply, pollution control, fisheries management, navigation, land drainage, and flood control.

The major thrust of postwar water policy in Great Britain has been to achieve the co-ordination of water use and development activities. In 1943, the Central Advisory Water Committee of that time (the Milne Committee) brought to public awareness for the first time the consequences of the prevailing "patchwork system" of river control and the need for co-ordination of water use and development functions, particularly in view of trends in population growth and concentration. The Milne Committee appraised the situation as follows:

> The boundaries of the larger local government areas necessarily bear no relation to watershed areas. There are in consequence many separate Authorities with responsibilities in relation to the same river basin . . . .

[24] It should be noted that the term "demand" tends to be used synonymously with "requirements" and/or "needs," which express a physical quantity based on a set of assumptions. It is seldom used in the sense of economic demand, which expresses a price-quantity relationship.

There is thus a patchwork system of control, with several Authorities . . . dealing with the one river, or, what is worse, neglect arising from the fact that with so many responsible Authorities, each Authority may be inclined to leave the duty of enforcing the Acts to others.

We are of [the] opinion, . . . that the principal defect of the existing system is not the overlapping functions, nor the possibility of conflict between interests, but the fact that no single body is charged with the duty of co-ordinating the various river interests or with the duty of ensuring that the requirements of all such interests are fully weighed when questions affecting the river are under review, with the result that the river is not used to the best advantage of all the interests concerned.[25]

The Committee's final report in 1943 recommended the creation of new comprehensive river authorities to administer the functions of river control and pollution abatement and to play a central role in gathering data and advising relative to water conservation for municipal and industrial uses.[26] Whether the River Boards, established in 1948, fully met the Milne Committee's concept of "new comprehensive river authorities" may be a matter for debate. The Water Act of 1945 and the River Boards Act of 1948, both directly following the Milne Committee Report, were, it would seem, largely concerned with two separate aspects of water co-ordination, with only slight legislative recognition of any functional relationship between the two.

The Water Act of 1945

The Water Act of 1945 gave to the Minister of Housing and Local Government a major responsibility in water resources policy, namely, "to promote the conservation and proper use of water resources and the provision of water supplies in England and Wales and to secure the effective execution by water undertakers . . . of a national policy relating to water."

[25] Central Advisory Water Committee, *Final Report* (1943), *op. cit.*, pp. 26–27.

[26] Reported in Central Advisory Water Committee, The Trade Effluents Sub-Committee (the Armer Committee), *Final Report* (HMSO, 1960).

The Act is significant to water resource policy in three ways. First, it established a clear interest and role for government in assuring adequate water supplies and an efficient water supply industry. Second, it recognized for the first time that government's interest in water supply derived from something more than its concern for public health. The Water Act of 1945 replaced the Public Health Act of 1936 as the basic code under which statutory water undertakers operate.[27] Third, the 1945 Water Act seems to express a commitment to centralization of governmental involvement in water supply, all of the new significant powers being assigned directly to the Minister.

Two kinds of functions were elaborated in the Act:

1. Measures for conservation of water for supply, to be exercised directly by the Minister of Housing and Local Government. Chief among these were:

(a) authority to license and control new or altered abstractions of ground water in areas where the public interest requires special measures for the conservation of water, in order to protect the water supplies for municipal, industrial or other purposes (Sec. 14);

(b) authority to protect against pollution of any water whether surface or underground which is developed and delivered to users by statutory water undertakers (Sec. 18–20);

(c) authority to require reports and records of the quantity and quality of water abstracted from any source, and of the location, nature, and operation of wells (Sec. 6, 7);

(d) authority to require local authorities and statutory water undertakers to carry out surveys of water requirements and supply potentials, and to formulate proposals for meeting future requirements in their respective service areas (Sec. 5).

2. Measures designed to promote efficiency of water supply services, consisting principally of authorities of the Minister to

[27] The new code is contained in Schedule 3 of the 1945 Act.

supervise and direct the statutory water undertakers. Chief among these were:

(a) authority to require water undertakers to supply water for non-domestic purposes upon request of users within their respective service areas, subject to priorities for domestic needs (Sec. 27);

(b) authority to require the combination of two or more water undertakings, or the establishment of other arrangement to better rationalize the pattern of water undertakings (Sec. 9);

(c) authority to delineate the area of supply that water undertakers may develop (Sec. 10) and authorize the supplying of water to customers outside an undertaker's service area (Sec. 11);

(d) authority to authorize new water supply undertakings and the extension and expansion of established ones (Sec. 23);

(e) authority to alter rates and charges which undertakers are authorized to levy (Sec. 40);

(f) authority to take over and operate a water supply undertaking if there is adequate evidence that an undertaker or local authority is not providing an adequate amount or level of water supply service (Sec. 13).

Perhaps most significant of the accomplishments under the 1945 Water Act has been the general supervision and direction to water undertakers and local authorities with respect to their water supply enterprises, particularly in rationalizing their service areas. Of a total of about 1,200 separate water undertakings that existed prior to the Act, consolidation left approximately 400 undertakings by 1963.[28]

[28] Report by the Joint Parliamentary Secretary, Ministry of Housing and Local Government (Lord Hastings) to the House of Lords in response to a Parliamentary question from Lord Wakefield of Kendal. Extract from proceedings in the House of Lords (undated) during 1964.

The River Boards Act of 1948

Under the River Boards Act of 1948, England and Wales were divided into thirty-two River Board districts, and the primary activities relating to river control and quality were placed in the River Boards. These new river agencies were direct descendants of the Catchment Boards, which were organized under the Land Drainage Act of 1930 to carry out and finance land drainage projects and related works for flood control in main river channels.

The 1948 Act broadened and strengthened the Catchment Boards. It provided the River Boards with two kinds of functions, those of previous statutes, which were transferred to the River Boards, and those functions newly authorized by the 1948 Act. The transferred functions gave to the Boards comprehensive authorities for the development not only of land drainage and flood control but of fisheries projects and, in certain districts, navigation and harbor projects; for the regulation of fisheries; and for the enforcement of anti-pollution legislation then in effect.[29] In addition, the 1948 Act:

1. imposed on each Board, in the exercise of its functions, the duty to conserve the water resources of its area;

2. authorized each Board to provide facilities for and collect basic water resource data and provide such information relating to water resources as the Minister of Housing and Local Government might request, including specifically information on abstractions and discharges in the River Board area;

3. authorized each Board to require information on abstractions and discharges from any person or enterprise who in the opinion of the Board is abstracting from natural water bodies in the River Board area.

Four observations about the River Boards are relevant:

1. The Boards were successful in integrating functions relating to drainage, flood control, fisheries, and pollution control

[29] Primarily, the Public Health Act of 1936, the Rivers (Prevention of Pollution) Act of 1876, and the Salmon and Freshwater Fisheries Act of 1923.

but took only halting steps toward co-ordination of these functions with activities related to water conservation, that is, water
availability for withdrawal uses.

2. The accomplishment of River Boards in water conservation was limited by deficiencies in their legal authority and the
institutional environment in which they were expected to
operate. Their principal specific authority relating to water
supply was to collect and report basic data regarding water availability. The law placed the burden of initiating reports from
abstractors upon the River Board, rather than making abstractors responsible directly. This proved to be difficult to administer since the Boards had little or no business with water undertakers or industries; therefore, they had no position of strength
from which to require reports.[30]

3. Pollution abatement activities were strengthened by virtue
of the fact that the Boards were given enforcement powers of
local authorities under the 1876 Prevention of Pollution Act
and of Fishery Boards operating under fishery laws. Enforcement by local authorities had been ineffective because, it was
alleged, local authorities are often the polluters. Responsibility
for fisheries laws gave the River Boards an independent interest
in quality that strengthened enforcement.

4. In spite of their greater success in pollution control, River
Boards still found their ability to act falling short of the needs.
In successive legislation, Rivers (Prevention of Pollution) Acts
of 1951 and 1961, practically every waste discharge was put
under licensing control. These laws replace the Rivers (Prevention of Pollution) Act of 1876 in its entirety as well as relevant
water pollution control provisions in the Public Health Acts of
1875 and 1936, as amended.

Policy and institutional setting in the 1950's

During the fifteen-year period that followed the River Boards
Act (1948–1963), water use and development had two centers of
public policy administration. The Ministry of Housing and

---

[30] In contrast, under the 1963 Water Resources Act, abstractors are required to be licensed by the relevant River Authority at which time data
regarding present and projected use are required.

Local Government was primarily concerned with water supply. Many of its efforts were directed to supervising and directing water undertakers and local authorities and to licensing ground water abstractions, as authorized by the 1945 Water Act. River Boards, under general management control of local governments, were river oriented and were dominated by land drainage, flood control, and fisheries functions. The Boards provided the primary focus for river basin integration of water development activities. Although the Minister of Housing and Local Government shared with the Agriculture Minister general responsibility for overseeing River Boards' operations, this was limited principally to broad policy and organizational and procedural matters prescribed in the Act. The Ministry of Agriculture had close day-to-day operating relationships with the Boards on drainage and flood control schemes and on questions of fisheries management. Consequently, Agriculture appears to have become the most influential Cabinet ministry in the affairs of River Boards.

In later years, however, the River Boards were brought into a closer working relationship with the Ministry of Housing and Local Government through two water resource activities: the issuance of permits for effluent discharges and the conduct of hydrologic surveys. Responsibility for administering discharge permits was assigned to River Boards by the Rivers (Prevention of Pollution) Acts of 1951 and 1961. However, the River Boards' powers under these Acts were subject to supervisory procedures prescribed by the Minister of Housing and Local Government. These supervisory activities involved the Ministry more directly in River Board operations than had been the case under previous pollution abatement laws.

Authority to launch a program of hydrologic surveys by the Minister of Housing and Local Government also contributed to closer operational relations of the ministry with River Boards. Although the Water Act of 1945 authorized the Minister to promote surveys of needs and the formulation of proposals for meeting needs, action in this area had been slow. The first report in 1959 of the Sub-Committee on the Growing Demand for Water gave renewed emphasis to the importance of such surveys and emphasized that they should involve comprehensive examination in each river basin of rainfall and run-off, public and private sources of supply, effluent discharges, reuse of water and potential storage sites. This report resulted in the Ministry of Housing and Local Government expanding and accelerating its surveys, including the development of comprehensive river basin reports. Expanding the scope of surveys appears also to have contributed to closer associations with River Boards than had previously existed.

This, then, is the policy and institutional setting that prevailed at the time of the development and enactment of the Water Resources Act of 1963.

The Water Resources Act of 1963—A Water Management Charter

The 1963 Act had its immediate origins in a series of three reports issued by a sub-committee of the Central Advisory Water Committee.[31] The Sub-Committee on the Growing Demand for Water was first appointed in October 1955 to consider: (1) the demand for domestic, industrial and agricultural water; (2) the problems in meeting the demand; (3) the costs; and (4) whether economies in the use or cost of water could be made.[32] The first report of the sub-committee, made in December 1958, pointed out future shortage areas and made some general recommendations primarily concerned with improving data and information. In general, however, the tone lacked urgency— perhaps because of its lack of specificity. The demand estimates made in the first report did not consider agricultural irrigation, and the parent committee asked the sub-committee to make a special study of irrigation to determine "the extent of uncontrolled abstractions of surface water for agriculture, especially irrigation, for industry and for other purposes . . .; [and] to

[31] A standing committee, advisory to the Minister of Housing and Local Government and appointed by him in accordance with the *Water Act of 1945*.

[32] Sub-Committee on the Growing Demand for Water, *First Report* (1959), *op. cit.*, p. 1.

consider whether powers are needed to control such abstractions in general or in particular; and to make recommendations."[33]

A severe drought was experienced in the crop season of 1959, and the sub-committee was prodded "to advise as a matter of urgency on the basic question of whether a need to control abstractions had been established."[34] In replying to this immediate question, its second report in 1960 concluded that potential irrigation demand was sufficient to require control of abstraction from surface waters. The sub-committee made clear the interim character of the second report and its intention to continue its studies and present "a more detailed report at a later date on such matters as the form of authority to operate any system of control and the method of operation."[35]

The third and final report of the sub-committee (the Proudman Committee Report) set forth in some detail proposals for a comprehensive water policy and unified administration of water use and development activities in river basin areas.[36] The government's White Paper[37] followed quickly and in general endorsed the Proudman Committee Report. Thus the ground had been laid for the drafting and passage of the Water Resources Act of 1963 with a minimum of controversy. The 1963 Act achieves the goal first expressed by the Milne Committee in 1943 and elaborated by the Proudman Committee in 1962, namely, a comprehensive legislative basis for integrated water management within groups of hydrologic units. The highlights of these twenty years of development are outlined in Table 1.

Summary of Water Management Institutions

The institutional structure provided in the 1963 Water Resources Act represents a fundamental integration of the two essentially separate thrusts that characterized evolution of water

[33] Central Advisory Water Committee, Sub-Committee on the Growing Demand for Water, *Second Report* (HMSO, 1960), p. 3.

[34] *Ibid.*

[35] *Ibid.*

[36] Sub-Committee on the Growing Demand for Water, *Final Report* (1962), *op. cit.*

[37] *Water Conservation, England and Wales,* Cmd. 1693 (HMSO, April 1962).

TABLE 1. HIGHLIGHTS IN THE EVOLUTION OF WATER POLICY IN ENGLAND
SINCE WORLD WAR II

Milne Committee Report, 1943

Recommended creation of river authorities to integrate activities relating to river control, water pollution, and water conservation. Denied need for additional pollution control powers.

Water Act, 1945

Gave the Minister responsibilities for water conservation, largely through reporting on abstractions, licensing of ground water withdrawals, and supervision of water undertakings—especially in rationalizing their service areas and sources of supply.

River Boards Act, 1948

Created River Boards to conduct activities relating to river control and pollution abatement with only permissive authority to collect information on water availability and use.

Hobday Committee Report, 1949

Recommended additional pollution control powers be given to River Boards; basis for Rivers (Prevention of Pollution) Act of 1951.

Rivers (Prevention of Pollution) Act, 1951

Reaffirmed policy prohibiting pollution; authorized River Boards to prescribe effluent standards as guide for enforcement of existing discharges; and required new or altered discharges to obtain "consent" of River Boards.

Armer Committee Report, 1960

Recommended that principle of "consent" procedure of 1951 Act be extended to pre-1951 discharges.

Rivers (Prevention of Pollution) Act, 1961

Required "consent" for pre-1951 discharges; directed periodic review of all consents with authority to change conditions or revoke permit; and extended "consent" procedures to estuaries and tidal waters.

Proudman Committee Report, 1962

Recommended creation of River Authorities with all powers of River Boards, plus comprehensive powers relating to water conservation, including resource surveys, planning, licensing of all withdrawals and charging therefor, and construction of water conservation facilities.

White Paper, 1962

Generally endorsed recommendations of Proudman Committee; modified membership of governing boards of River Authorities.

Water Resources Act, 1963

Enacted recommendations of Proudman Committee as modified by White Paper, to come to full effect April 1, 1965.

resource policy in England since the war. In a sense, the River Authorities are the vehicle of integration. They now possess by transfer: (1) the power of the River Boards to control the behavior and quality of surface waters, including authorities relating to drainage, flood control, fisheries, pollution control, and in some cases, navigation, and (2) the initiative power of the Minister of Housing and Local Government under the Water Act of 1945 for the conservation and proper use of water resources. In addition, River Authorities are given an impressive scope of new powers and responsibilities. First, and most important, they are given the general duty "to take such action as they may from time to time consider necessary or expedient, or as they may be directed to take by virtue of this Act for the purpose of conserving, re-distributing or otherwise augmenting water resources in their area, or securing the proper use of water resources in their area, or of transferring any such resources to the area of another river authority."[38]

Further, the Water Resources Act of 1963 empowers the River Authorities to exercise specific new authorities significant to their role as water management agencies. For example, they are authorized and directed to: (1) construct and operate a comprehensive network for the collection of hydrologic data, (2) conduct surveys and formulate proposals for action in the River Authority's area, (3) establish "minimum acceptable flows," (4) license all water withdrawals, (5) levy abstraction charges, and (6) construct, operate and co-ordinate reservoirs and other facilities that control, or make accessible, water resources.

The allocation of governmental powers and functions by the Water Resources Act of 1963 results in a highly decentralized system of water management. The River Authorities are the keystone of the system. It is in these twenty-seven semi-independent authorities that the principal water management actions are initiated. It is in these agencies that the basic conflicts implicit in water management decisions are first faced. In addition, three major units in central government play significant roles by articulating national interest in regard to water management

[38] *Water Resources Act of 1963, op. cit.*, Sec. 4.

decisions and provide both general and specific supervisory control over River Authorities. These central authorities are: (1) a newly established Water Resources Board, which is the key agency of central government for national water resource planning; (2) the Minister of Housing and Local Government, who is principally responsible for determinations regarding the government's supervision and/or control of statutory water undertakers, local authorities and industry with respect to water supply and pollution control; and (3) the Minister of Agriculture, Fisheries and Food, who continues to exercise the central powers regarding drainage, flood control, and fisheries essentially as he did with the River Boards. The two Ministers share general supervision of River Authorities, particularly with respect to matters relating to their establishment, membership, and general operating policies.

A fuller treatment of the institutional "system" is presented in Chapters 4 and 5.

4

GOVERNMENTAL POWERS FOR WATER MANAGEMENT IN ENGLAND

The Water Resources Act of 1963 integrates a wide scope of legal powers and administrative activities for the purpose of controlling and/or influencing both water use and development.

Three major institutional features of the water management system for England and Wales are elaborated in the 1963 Act. These are: (1) the legal powers given to government to intervene in water resource use and development, (2) the organizational arrangements through which those authorizations are administered, and (3) the means of financing water management activities. This chapter will examine the substantive legal powers for water management, while Chapter 5 will provide a descriptive analysis of both the organizational structure and the financial arrangements.

British law provides six types of legal powers relating to water use and development that serve as the primary tools of water management. Some of these basic legal powers are new or have been expanded by the Water Resources Act of 1963, but many were enacted by earlier statutes and became integrated into the water management system by the 1963 Act. The six types of legal authorizations are: (1) collection of information and proposal for action; (2) determination of "minimum acceptable flows"; (3) licensing discharge of waste water; (4) licensing water withdrawals; (5) levying abstraction charges; (6) construction and co-ordination of water development. In examining the origins and significance of these powers, three objectives are sought: first, to provide a brief but comprehensive review of the historical evolution of governmental involvement in each type of activity; second, to summarize the nature and extent of present administrative authority in each field of action; and third, to describe the administrative matrix, that is, the procedural relationships among agencies that are required in the administration of the several kinds of water management measures.

Collection of Information and Proposals for Action

In general, prior to the 1963 Water Resources Act, legislative authorizations for collecting information and developing action programs were inadequate, and when provided were scattered, thus making co-ordination of regional water surveys and plans difficult, if not impossible. Until the creation of River Authorities, little attention had been given to a comprehensive, systematic, and continuing assessment of water resources and needs on a regional basis. Even the collection of elemental hydrologic data had lagged far behind need.

Historical roots

A brief review of relevant legislation indicates the slow development of this aspect of water management. A Surface Water Survey was established under the 1936 Public Health Act. Since the creation of the Ministry of Housing and Local Government, the Survey has operated in that ministry. One of its principal functions has been the publication of the *Surface Water Yearbook*, which, in 1964, was based upon reports from over 270 stations in England and Wales.

The Water Act of 1945 gave to the Minister[1] power (1) to require both public and private water undertakers to carry out surveys of the demand-supply situations in their service areas and formulate proposals for meeting future needs, and (2) to require persons withdrawing water from natural water bodies to furnish records and information. Power of the Minister under this Act to license abstractions of underground water strengthened his role in collecting data on ground water use in those areas where the licensing power was exercised. In addition,

[1] The Minister, unless otherwise specifically designated, means the Minister of Housing and Local Government.

the Geological Survey accelerated and expanded ground water studies, particularly in the "Conservation Areas," which are subject to abstraction control under the 1945 Act.

In 1948, River Boards were authorized to develop a scheme for measuring rainfall, runoff, and river flows in their respective areas and were encouraged to do so by grants-in-aid for the installation of hydrometric schemes, which were to be provided through the Ministry of Agriculture, Food and Fisheries. Additionally, the Boards were authorized to require reports from major abstractors of both surface and underground water in the River Board areas, and the Boards were required to collect hydrologic and water use data as directed by the Minister.

In 1964, at the time this study was started in England, there seemed to be general consensus that none of these powers to collect hydrologic and water use data had been used as fully as the language of the law would have permitted. However, the Surface Water Survey had been given some new impetus by the first report of the Sub-Committee on the Growing Demand for Water in 1959.[2] This report recommended that hydrological surveys should be undertaken in regions where considerable pressure upon the water resources is evident. It was decided that such surveys should comprise a comprehensive examination in each river basin of rainfall and runoff, public and private sources of supply, effluent discharges, reuse of water and potential storage sites.[3] The activities of the Surface Water Survey were expanded to carry out this recommendation. At the time of this study, a basin-wide survey had been completed for the River Severn and work was underway on several other river basins. The Severn report was considered by the Surface Water Survey to be a prototype for water surveys of the future.

Present scope and character

Prior to the Water Resources Act of 1963, not only were the legal authorities relating to the collection of basic hydrologic

[2] Central Advisory Water Committee, Sub-Committee on the Growing Demand for Water, *First Report* (HMSO, 1959).
[3] Memorandum, "Hydrologic Surveys—Explanatory Note," from A. Gerald Boulton, Engineer in Charge, Surface Water Survey, one page, mimeographed, March 1962.

data highly fragmented, but there was little recognition of the need for planning development projects and other actions designed to enlarge the capacity of water resources to provide water services. In contrast, the 1963 Act authorizes an integrated function for data collection, resource surveys, and development planning. The new Water Resources Board and the River Authorities are the key elements in performing the function. They are given co-operative and mutually supportive functions regarding the collection of information and the formulation of plans for action, which are prescribed in some detail in the statute. The Water Resources Board, in addition to providing the national perspective on resource assessment and development plans, exercises a considerable leadership role over the River Authorities in their survey and planning activities.[4]

The River Authorities are *required* to perform three types of activities:

1. Planning, installation and operation (upon approval of Water Resources Board) of a hydrometric scheme for measuring and recording rainfall, evaporation, flow (level or volume) of inland waters (surface and ground waters) and other matters which are likely to affect water resources in the area; registering the installation and operation of any hydrologic measuring device in River Authority area not under operation by the Authority; and reviewing the adequacy of the data collecting scheme periodically, at intervals of not more than seven years.

2. Surveying the water resources of the area, including estimates of demand over a twenty-year period and identification of supply problems; formulating proposals for action by the River Authority; and reviewing and revising at intervals of not more than seven years such surveys and proposals, giving regard to the advice of the Water Resources Board with respect to national and interregional considerations.

3. Making data, records, and reports available to local authorities, water undertakers, industry, and the general public; and

---

[4] For the reader seeking a more thorough understanding of the reciprocal relations between River Authorities and the Water Resources Board, reference should be made to Chapters 3 and 5 of this study.

reporting data, surveys, and proposals to the Water Resources Board and appropriate ministers.

In discharging its responsibility for the national perspective, the Water Resources Board is directed to conduct a continuing surveillance of need "for conserving, redistributing and augmenting water resources." The Board's responsibilities for nationwide assessments is interlinked with the resource assessment and planning functions of the River Authority. These are reflected both implicitly and explicitly in the following functions that are assigned to the Board:

1. Advising and assisting River Authorities in preparing hydrometric schemes, making water resource surveys, and formulating proposals.

2. Reviewing and approving hydrometric schemes of River Authorities, thus qualifying schemes for central government grants to cover construction costs.

3. Conducting such research, making such inquiries, and submitting such reports as the Board may consider necessary or expedient, or as a Minister may require; and directing River Authorities to supply specific data and/or to prepare special reports as may be required by the Board.

4. Collating and publishing information relating to water resources and demand thereupon.

5. Considering action needed for the purpose of conserving, redistributing, or otherwise augmenting water resources, and recommending actions to Ministers and to River Authorities.

To aid the Water Resources Board in carrying out these functions, the Surface Water Survey was transferred from the Ministry of Housing and Local Government and selected personnel from the ground water department of the Geological Survey. These two professional groups have been made separate organizational units under the Board.[5]

---

[5] The Board has announced that its organization consists of five principal divisions: (1) Geology, (2) Hydrometry, (3) Liaison and Promotion (with Agriculture, Industry, River Authorities, and Public Water Suppliers), (4) Planning, and (5) Research. (*Second Annual Report of the Water Resources Board* for the year ending September 30, 1965.)

The institutional arrangements provided by the Water Resources Act make possible a fruitful conjunctive operation of River Authorities and the Water Resources Board in developing national and sub-national policies and plans. The emerging operational relationship may be illustrated by the recently completed report, *Water Supplies in Southeast England*.[6] The Water Resources Board describes the report as dealing with ten River Authority areas in Southeast England from the Wash to Dorset. "Together these areas comprise a region with smaller indigenous water resources than any other part of the country, but with a heavy and rapidly growing demand *and it has become clear that its water supply problems need looking at as a whole*."[7] (italics supplied) After describing the various ways the resources of this important component of the nation can be augmented, the Board outlined a two-part program: (1) immediate action to secure the region's position for the next decade; and (2) further research and investigation as a basis for determining strategy for the years thereafter.

In the introduction to its report, the Board acknowledges the help of the ten River Authorities in the study area as well as of corresponding area committees of the British Waterworks Association. The report went on to say:

> The river authorities have the responsibility under Section 14 of the Water Resources Act 1963 for preparing surveys of the resources of their areas and plans to meet future demands for water. Heavily engaged as they have been with the task of licensing existing abstractions the river authorities have not yet been able to carry out these surveys. Meanwhile, we [The Water Resources Board] have a duty under Section 12 of the Act to consider what action needs to be taken . . . and to advise the Minister and the river authorities. The needs of the South East are too pressing to permit us to await the Section 14 proposals of the river authorities, although this would have made our task much easier. Indeed, the river authorities will need some guidance from us before they can

---

[6] Water Resources Board, *Water Supplies in Southeast England* (HMSO, July 1966).

[7] *Third Annual Report of the Water Resources Board* for the year ending September 30, 1966 (HMSO, December 1966), p. 15.

formulate their own proposals [plans for action] because the problems go beyond their separate areas. . . . *We hope that this report will provide the necessary regional framework within which the river authorities can plan.[8]* (italics supplied)

Here then there is emerging an operational relationship in which the River Authorities, in addition to assembling data and conducting surveys to support their own planning, contribute intelligence inputs to policy and strategy studies of the Water Resources Board, which in turn provide a "regional framework" within which River Authorities can better plan. A similar regional report has been prepared on the water resources in the North. These reports comprehend a larger demographic-economic region than that of a single river basin and they emphasize water supply rather than hydrology. This kind of survey will undoubtedly become more common and will supplement, if not replace, the resource inventory emphasis which characterizes the Severn model.[9]

Clearly, the Water Resources Act of 1963 provides authority and structure for a more unified and purposeful system for the assessment of water resources and the formulation of action programs than had previously existed. In a very large degree, the establishment of the Water Resources Board is the key to the difference. As a new agency, it has been given responsibility for leadership, co-ordination, and follow-through on data collection, resource surveys, and planning, the previous lack of which appears to account for the failure to exercise specific legislative authorizations. In addition, under the 1963 Act, the River Authorities are in a better position than the River Boards under the 1948 Act to perform data collection, assessment, and planning functions. Four points in this regard seem relevant:

1. The River Authorities are now responsible for water conservation and thus are properly parties to water supply information, whereas River Boards had no functional connection with water undertakers, even though they had a legal authorization to collect information from them.

---

[8] *Water Supplies in Southeast England, op. cit.*, pp. 1–2.
[9] See p. 49.

2. The River Authorities will now be in possession of comprehensive information on withdrawal uses of water through the administration of their powers to license abstractions.

3. The River Authorities are required within one year to submit to the Water Resources Board a proposed hydrometric scheme, which, if approved, the Board will support for a government construction grant. The River Boards Act of 1948 said, on the other hand, that the Boards *may* submit schemes for the measurement and recording of rainfall, etc. and *may* receive financial support under a drainage act administered by the Minister of Agriculture, which act required qualifying procedures, often prohibitive to River Boards.

4. The River Authorities are given a *duty*, as soon as practicable after their organization, to conduct a survey of the water resources in their respective areas and to formulate proposals (plans) for action. The River Boards had no such comprehensive survey and planning authority; they only had a collection of single-purpose planning authorities associated with functions transferred to them in the fields of drainage, flood control, fisheries, navigation (in some cases), and pollution control.

Together, the Water Resources Board and the River Authorities possess more explicit authority and provide a more unified institutional system for the collection of water resource data, the survey of needs, and the planning of action than had previously existed.

Determination of "Minimum Acceptable Flow"

Closely related to the authority for surveys and plans is the new statutory requirement that each River Authority determine the minimum acceptable flow (or level in case of a lake or aquifer) at critical points throughout its area.

Minimum acceptable flow (MAF) is described as " . . . the minimum flow, level or volume, at a specified control point, which—having regard to the character of the water and its surroundings (including natural beauty, if any) and the water flow from time to time—is in the opinion of the River Authority needed for safeguarding the public health and for meeting, both quantitatively and qualitatively, the requirements of existing lawful uses. . . ."[10]

The requirement to set MAF levels has been a subject of considerable discussion in England. Following the passage of the 1963 Act there seemed to be no clear agreement as to the intended role of MAF in water management and several different interpretations emerged in planning the implementation of the Act.[11]

The evolution of the concept

The legislative history suggests that there was a wide range of opinion among the framers of the 1963 Act as to the way in which MAF was to be used. At one extreme, the determination of an acceptable low flow seemed to be viewed solely as establishing the condition necessary and sufficient to prohibit new abstractions, that is, withdrawals by persons and for purposes not considered by the law to have a "license of right" to water.[12] Proponents of the other extreme seemed to view MAF as a level or flow which should serve as the goal for comprehensive water development schemes. Under the latter, there is a presumption that if a higher "minimum" than the probability of nature promises is considered desirable, development efforts will be directed to providing storage (including, presumably, by aquifer recharge) and flow regulation to bring "minimum" flows up to some level of agreed-upon needs. The law itself is not clear as to how MAF determinations are to be used or how they relate to other governmental measures for water management.

Basically, the concept embodied in MAF derives from the long standing requirement for compensation water to be released from supply reservoirs operated by private and public

[10] D. E. Tucker, "New Concepts in Water Conservation," *Journal of the Institution of Public Health Engineers,* October 1965, p. 286.

[11] It should be noted that the question of how MAF should be expressed and its dependability is a difficult one. Since streamflows are stochastic variables, it may be misleading to establish MAF in absolute terms when, in fact, the significance of flow levels is found in the frequency with which each level may be expected. It has not been possible within the scope of this study to determine to what extent this problem has been considered in conceiving MAF as a central water management measure in England.

[12] See pp. 71 ff.

water undertakers. Theoretically, the amount required was set at a level necessary to protect existing riparian rights of down-stream users. This approach seemed to be based upon the assumption that the rights of riparians were to specific quantities and that any fraction of flow above that was in a sense "surplus" and could be stored and allocated to non-riparians.

When concern about water conservation began to grow, the requirement for compensation water came under attack, because its rigidity prohibited a full use of the water resource during its passage to the sea. Compensation flows were usually based upon average flows. This resulted in releasing water at times when it was not particularly necessary, while at other times water was needed downstream in greater quantities than compensation amounts required. The Proudman Committee in its report[13] gave considerable attention to this problem and recommended the MAF procedure essentially as provided in the law. But the Committee's report, like the law, is ambiguous as to how MAF might serve as a "tool" of water management.

There are indications that the ambiguity of the report reflects compromises, but not complete reconciliations, between two groups in the Committee—those who in meeting future shortages would put primary reliance on prohibiting "low priority" uses, and those who would favor a new positive policy where emphasis would be given to planning and executing developments designed to augment the capacity of a given stream. The latter group tended to conceive MAF as serving as a standard and guide for the development program, *as well as* a guide in the administration of abstraction licenses.

The role of MAF in water management

The government's White Paper on the Proudman Committee's recommendations is quite explicit about the dual role that MAF was expected to play. It said, "When the scheme (specifying minimum acceptable flows) has been approved by the Minister . . . , it will guide the authority in their control of

[13] Central Advisory Water Committee, Sub-Committee on the Growing Demand for Water (the Proudman Committee), *Final Report* (HMSO, 1962).

surface water abstractions and in considering the need for works to make more water available."[14] The Water Resources Act, however, is not as clear on this point. In fact, the specific provisions relating to setting MAF levels emphasize that the determinations shall be based upon "the requirements of existing lawful uses." The emphasis upon existing uses might suggest that the purpose of MAF is to set a ceiling for abstraction licenses rather than to set a goal for development of water.

Under a comprehensive view of water management, it might be argued that if abstraction licensing is to assure the licensees the amount of water they have been authorized to withdraw, either of three alternatives are open to the water management agent. First, it might not issue licenses for amounts beyond the established MAF; or second, it could discriminate in issuing licenses, attaching to some a condition which automatically suspends the right to withdraw water whenever water flow or level falls to or below MAF; or third, it might undertake the construction of conservation impoundments, or other facilities, designed and operated to raise the minimum flow probability of the hydrologic unit subject to management. Which of these approaches would be followed in the administration of MAF in England was not clear when the Water Resources Act came into effect.

Other provisions of the Act, however, make it clear that water management policy in England will no longer depend upon, or give primary emphasis to, control of water use. The Act includes new powers designed to provide comprehensive and integrated development works as well as improved regulatory measures; and it seems clear that the government supports the more positive developmental thrust.

Administration of MAF

The establishment of minimum acceptable flows seems likely to occupy a central place in British water management. The Water Resources Board, in providing policy guidance to the River Authorities, recognizes three broad but distinct groups of

[14] *Water Conservation, England and Wales,* Cmd. 1693 (HMSO, 1962), p. 6.

water management provisions: (1) intelligence and planning, (2) control, and (3) execution (of developmental works).[15] The determination of MAF is the bridge between intelligence and planning, and control. MAF provides the standard for "control" operations, particularly abstraction licenses and discharge consents. If it is also to establish the physical goal for which water development facilities are constructed and operated, it will also provide the bridge to the "execution" functions recognized by the Water Resources Board. MAF may well serve as a kind of operational cement to the whole water management job. Certainly, the determination of MAF may become one of the most critical and sensitive decisions the institutional system has to make; thus the allocation of responsibility for taking this action is of special interest.

It is significant that law *requires* each River Authority to initiate MAF determinations "as soon as practicable." The administrative procedure required may in general be outlined in five steps.

1. The River Authority and the Water Resources Board co-operatively decide for which waters in the Authority's area MAF determinations will be made. If they cannot agree, the Minister will decide and *direct* the Authority accordingly.

2. The River Authority prepares and submits to the Minister proposed MAF determinations.

(a) Such proposals indicate control points at which measurements are made and a recommended MAF for each point.

(b) In determining MAF, the River Authority is directed (i) to consider the character of the water and its surroundings, including particularly its natural beauty, and (ii) to specify a flow which shall be not less than the minimum necessary for safeguarding the public health and for meeting both quantity and quality requirements of existing lawful uses.[16]

---

[15] Water Resources Board, Memorandum No. 1, February 23, 1965.

[16] Section 19 (5) of the Act says "existing lawful uses . . . whether for agriculture, industry, water supply or other purposes, and the requirements of land drainage, navigation and fisheries."

(c) In preparing MAF proposals, the River Authority is directed to consult with relevant private and public water undertakers, internal drainage boards, navigation authorities, harbor authorities, and conservancy authorities, and where appropriate with the Minister of Transport and the Central Electricity Generating Board.

3. A River Authority proposal for MAF is to be made available for public inspection for twenty-eight days prior to its being submitted to the Minister. Notice of its availability must be duly published, and the public is invited to report objections to the Minister.

4. The Minister is responsible for approval of MAF determinations, thus giving them the force of administrative law. If the Minister does not approve, with or without modifications, he may direct the Water Resources Board to prepare a substitute proposal, or he may, in fact, prepare his own proposal. In either alternative, proposals are subject to the same rules of preparation and public review as those imposed upon the River Authority.

5. The River Authority is required to give continuing review to its MAF levels, and at intervals of not more than seven years to prepare and submit a substitute proposal (in accordance with the same procedures), including any such amendments deemed desirable.

Authority to set minimum acceptable flows is new and untried. How important it will become as a governmental measure for the direction of water management in England will largely depend upon the leadership of the River Authorities and of the Water Resources Board. In actual practice, MAF could provide a standard or a framework within which other water management decisions are guided. The establishment of MAF, therefore, may make possible a greater delegation of responsibility for other water management actions than would otherwise be allowable. MAF determinations implicitly subsume decisions about the uses and users that will be satisfied, and whether by restrictions on use or by construction of new developments. These are, of course, fundamental decisions and are shared by the River Authority, the Water Resouces Board, and the Minister. But, as in the case of most standards, MAF determinations may also tend to reduce the flexibility of River Authority decisions to meet changing circumstances from place to place, and from time to time. The effectiveness of standards, such as MAF, depends upon a delicate balance between the desirability of providing policy guidelines for decentralized decisions and the desirability of a high degree of flexibility in choosing from a range of possible courses of action at the time each decision is made.

Licensing Discharge of Waste

The Water Resources Act of 1963 provides little or no new substantive powers regarding water pollution control. The major sources of legal powers in this field, including administrative requirements, are the Rivers (Prevention of Pollution) Acts of 1951 and of 1961. Most of the administrative powers under these acts were assigned to the then existing River Boards. The 1963 statute transferred these powers to the new River Authorities. Today, therefore, the principal tool of water quality control is the legal requirement of the 1951 and 1961 Acts that anyone discharging effluent to natural waters (surface and underground) must obtain "consent" from the relevant River Authority. In practice, the River Authority receives applications for consent and issues a permit setting forth conditions to its consent. The present comprehensive system of consent permits was adopted after nearly a century of experimentation with other approaches to pollution control. Since water quality maintenance is one of the most critical aspects of water management, the background of government's involvement in pollution control in England deserves study in some detail.

Evolution of pollution control

The first effort to deal with water pollution in England came in 1876 in response to epidemics that plagued the nation in mid-nineteenth century. A new rivers act of that year known as the Rivers (Prevention of Pollution) Act was passed and served as the basic pollution control statute until 1951. The 1876 Act declared it illegal to put into any river or stream solid refuse, rubbish, sewage, or poisonous or noxious industrial wastes.[17]

In spite of the legislative prohibition, Turing reported nearly seventy-five years later that ". . . there are still towns, and not always small ones either, which empty their crude sewage straight into a river without any attempt at treatment whatever."[18]

The only enforcement powers in the 1876 Act were those given to local authorities to prosecute offenders. But since local authorities are often the polluters, and court costs were to be borne by the complainant, there were many counterpressures and few prosecutions resulted. Another basic weakness of the Act was its failure to provide any criteria of pollution. This left the courts with no guidelines for determining violations. Another deficiency was the fact that the Act provided no penalty for offenders, but only authority for the court to issue a stop and desist order. Such procedure had little relevance to the fact that with industrialization, the most serious pollution damage often resulted from a single, major industrial discharge. These kinds of "spills" from industry were particularly devastating to fisheries.

Fisheries interests gave organized leadership to the pollution fight, not just because of the increasing damage from industrial "spills" but also, according to Turing, because the Ministry of Health in carrying out the 1876 Act insisted upon limiting its concern to the health consequences of water pollution.[19] In 1923, the Salmon and Freshwater Fisheries Act introduced the idea of biological factors in water quality. It made illegal the placing into rivers of any matters ". . . to such an extent as to cause the waters to be poisonous or injurious to fish, or the spawning grounds, spawn or food fish."[20] This provision estab-

[17] The only exception being discharges established before 1876 if the discharger could prove to the courts that he was using the "best practicable and available means to render it harmless."

[18] H. D. Turing, *River Pollution*, the Buckland Lectures for 1950 (London: Edward Arnold and Co.).

[19] Standards adequate for health often may be lower than the biological requirements to maintain a full range of aquatic life. Water may be lifeless and still of no danger to public health.

[20] Turing, *op. cit.*

lished a more precise criterion of pollution—one at least that could be tested by relatively simple technologies. The Act also provided for a series of Fishery Boards whose jurisdictions covered specified river basins. These Boards were given not only prosecution powers for the purpose of this Act, but they were also given those powers in the 1876 statute which until the 1923 Act were exercisable only by local authorities. Thus armed, the Fishery Boards were more vigorous in prosecuting violators than the local authorities had been heretofore.

Concern over the lack of criteria of pollution led to early efforts to establish general standards. A Royal Commission was established in 1898 to study water pollution. One of its later reports (1912) recommended standards that would provide a guide for enforcement of the pollution control laws, that is, basically the Act of 1876.[21] The Commission's standards have had far-reaching effect on pollution control administration even though, with few specific exceptions, they have not been enacted into law nor made a part of regulations having the force of law. However, they are referred to frequently in any discussion of pollution control, and it is reported that most River Boards (1948–1965) had informally adopted them as working guides.[22]

When set up under the River Boards Act of 1948, it was believed that River Boards would provide a new instrument for pollution control. Many legislative authorities relating to rivers were transferred to the Boards, including power to prosecute polluters under the original pollution prevention act of 1876 as well as under the 1923 fisheries act. In addition, the River Boards were empowered to enter on property and to take effluent samples, which, if taken as prescribed in the Act, were de-

[21] The Commission's report set forth five classes of water in terms of biochemical oxygen demand and "expressed the view that if the biochemical oxygen demand (BOD) of a sewage effluent did not exceed 20 parts per million (ppm) and 30 ppm of suspended solids and if the receiving stream provided a dilution of 8 volumes of clean water to one of sewage effluent, the condition of the stream below the sewage works would be satisfactory." (From W. F. Lester, "Pollution Prevention," Symposium on River Pollution, November 12, 1957, *Journal of the Institution of Public Health Engineers*, April 1958.)

[22] Lester, *op. cit.*, pp. 67–71.

clared admissible as evidence in any legal proceedings regarding pollution. Assignment to the River Boards of responsibility for pollution control was characterized as "the greatest forward step in pollution prevention."[23] For the first time it made one authority responsible for a complete river system, rather than leaving the administration of pollution prevention in the hands of many local authorities. Efforts in 1948 to give the Boards stronger regulatory powers were countered by the argument that River Boards should be given a chance to demonstrate their effectiveness with the basic enforcement procedures already available to them. This position was advanced by the Federation of British Industries and the National Coal Board.[24] However, enforcement by court action proved increasingly inadequate, and the anti-pollution interests continued to press for more controls.

Introduction of discharge permits

Strong representations were made to the Rivers Pollution Prevention Sub-Committee (the Hobday Committee) in favor of a stringent system of licensing all waste discharges. Following that Committee's recommendations in 1949,[25] a new Rivers (Prevention of Pollution) Act was passed in 1951, which completely replaced the original enactment of 1876.

The 1951 Act, although restating the basic policy of prohibition which characterized the earlier act, is important for two innovations. First, it authorized the River Boards to adopt "byelaws" establishing effluent standards as a basis for improving control over existing discharges, and second, it required any one contemplating a new discharge, or a substantial alteration to an old discharge, to obtain consent from the appropriate River Board. Each permit to discharge may prescribe the conditions of discharge, including location, character, and capacity of the outlet, and the condition of the effluent. The conditions attached to each permit are recorded in a register maintained and

[23] Lester, *op. cit.*
[24] Central Advisory Water Committee, *Report of the Rivers Pollution Prevention Sub-Committee* (HMSO, 1949).
[25] *Ibid.*

made available to the general public by the River Board. Violators of discharge permits are subject to a single fine not to exceed £200.[26]

One of the key officials in the administration of the 1951 pollution prevention law reported in 1960 that the discharge consent procedure provided by the Act "has undoubtedly been most successful in controlling new discharges," but that the authority to set effluent standards "has proved most difficult and the prospects of comprehensive ones in the near future look bleak indeed."[27] A study committee reported in 1960 that no legal byelaw standards had been established under the 1951 Act.[28] However, it should be noted as was pointed out above, most River Boards were guided informally by the Royal Commission's standards of 1912. Moreover, working standards were announced in the annual reports of most River Boards and were used as the basis for the administration of the 1951 and 1961 Acts by the respective Boards even though the standards were never given legal status as River Board byelaws.[29]

Why the River Boards failed to exercise their newly found authority to adopt water quality standards is not clear. Viewing this kind of nonfeasance in the American context, one is tempted to attribute it to the inability of a local agency such as a River Board (and perhaps even now, the new River Authorities) to withstand economic pressures operating on the local political process. One authority on pollution in England, however, disputes the notion that failure to establish standards reflected a weakness of the River Boards.[30] There is some evi-

[26] Approximately $500.00.

[27] A. Key, "River Pollution and its Control, Present and Future," *Proceedings* of the Annual Conference of the River Boards Association at Porthcawl, 1960, p. 10.

[28] Central Advisory Water Committee, The Trade Effluents Sub-Committee (the Armer Committee), *Final Report* (HMSO, 1960).

[29] Correspondence with Dr. Louis Klein.

[30] ". . . it is not strictly true to say that none of the Boards availed themselves of this authority (to set byelaw standards). They certainly tried. I can myself recall many meetings with the Ministry of Housing and Local Government in connection with two rivers. . . . These meetings got nowhere, however, as the Ministry seemed unwilling to agree to the byelaw standards although we made many concessions to the manufacturers." (Author's correspondence.)

dence, particularly among the professionals in the River Boards, that objection to standards is based upon their relative inflexibility. Effluent standards, for example, do not allow for consideration of different conditions of receiving waters—either for differences in place or in time.[31] A former pollution control officer of the Trent River Board has stated, "Even if the many technical difficulties of making . . . standards are overcome, they would not possess the same flexibility as the legal standards laid down in consent conditions, which are tailor made to fit a particular discharge."[32]

Adoption of comprehensive discharge consents

Need for better control over existing discharges led the 1960 study committee to recommend that a consent procedure similar to that in the 1951 law be applied to pre-1951 discharges as well.[33] This recommendation was enacted in the Rivers (Prevention of Pollution) Act of 1961. Accordingly, since July 1963, the effective date of that Act, each River Board (now River Authority) has the duty and power to license *every* discharge into the natural waters of the area over which it has jurisdiction. The 1961 Act strengthens penalties on violators by setting no limit on fines that may be imposed by the Courts upon a conviction upon an indictment, but limiting the fine to 100 pounds on a summary conviction. Section 5 of the 1961 Act requires periodic review of any conditions attached to discharge consents issued under either act and authorizes making any reasonable variations or revocations of the permit. This provision at least permits the licensing system to maintain some flexibility. The discharge consent powers of the 1951 and 1961 statutes give each River Authority a comprehensive system of licensing discharges, thus directly controlling all major sources of pollution.

The Water Resources Act in 1963 provided little new sub-

[31] Although the Royal Commission's effluent standards were based upon a prescribed minimum factor of dilution by the receiving waters, there seems to be apprehension that this part of the standard may be overlooked.

[32] Lester, *op. cit.*, pp. 67–71.

[33] The Trade Effluents Sub-Committee (the Armer Committee), *Final Report*, 1960, *op. cit.*

stantive power for discharge control.[34] Its main contribution to better pollution prevention, according to one official, is the fact that it placed conservation and water supply responsibilities in River Authorities in association with the administration of discharge permits. Having these two responsibilities in the same administrative hands should provide greater motivation for rigorous application of the powers to license discharges, now centralized in the River Authorities.

Administration of discharge consents

The responsibility to issue discharge permits imposes upon the River Authority one of its largest administrative tasks, and may over time be the acid test of its decision making capability. The 1961 Act, unlike the 1951 law, brought upon the River Boards a flood of applications for consent that could not wait for study and deliberation. Applicants were already discharging from ongoing operations, and they could not immediately cease and desist until corrective action could be taken without serious consequences to the local and national economy. The 1961 statute provides a procedure whereby any polluter who has made application under the Act, thus supplying detailed information about his discharging effluent, will not be held in violation until further action is taken by the River Authority.

In the meantime, the more serious cases are processed first. A series of hearings and negotiations are held with the major polluters, with the objective of arriving at a mutually agreeable set of conditions which will attach to the discharge license. Polluters have the right to appeal consent decisions to the Minister of Housing and Local Government. The consent procedure may be a protracted one, particularly for the most serious cases. In the meantime, the other polluters continue their past practices until the River Authority can act, but at least the Authorities have, through the file of applications, a complete record of all discharges in their respective area, and this alone is a large step forward.

[34] The principal exception being that discharges of effluents into underground strata were put under regulation of the Rivers (Prevention of Pollution) Acts of 1951 and 1961.

Although the River Authority is now the primary administrative agent in the implementation of the 1951 and 1961 Acts, its consent actions are subject to both general and specific checks and balances. River Authority actions on applications for discharge consent may be appealed to the Minister who may rescind or modify the actions in accordance with his judgment of the merits of the case. Although the River Authority has wide discretion in periodically reviewing and modifying earlier consent decisions, the Minister may at any time direct a River Authority to vary or revoke a discharge permit. The Water Resources Board has little specific responsibility in the administration of discharge consents. However, under its several general functions set forth in Section 12 of the 1963 Act, the Board is required "to bring to the notice of the river authority concerned any case where it appears to the Board that, . . . the quality of the water contained in an inland water in that [River Authority] area needs to be improved, and that the requisite improvement could be obtained through the exercise of powers conferred by the Rivers (Prevention of Pollution) Acts of 1951 and 1961."

The Water Resources Board may also have considerable influence upon a River Authority's administration of discharge consents through the Board's general role of advising the Minister in the whole range of his responsibilities in water conservation.

Licensing Water Withdrawals

In areas where withdrawal of water threatens to limit other uses, a system of licensing withdrawals becomes an important tool for rational water management. The Water Resources Act of 1963 establishes for the first time in England a comprehensive system of licensing withdrawals from surface and ground water sources.

Evolution of withdrawal regulations

In England, the only control over abstractions, until after World War II, was either (a) that exercised under riparian law through damage claims sought in a court of justice, or (b) that effected by the assignment of "rights" by Royal and/or Parliamentary Charter for water undertakers, industries and navigation. These early legal constraints on water withdrawals gave no effective "management" control. Rights were granted without knowledge of the extent of the supplies that were to be drawn upon. Diversions from one drainage to another were authorized with little regard to future needs in the basin of origin. Over time, the resulting web of independent abstractors had become an intricate, competitive, and inefficient system. There was a growing need to rationalize supply sources and abstraction requirements. The first step in providing for such rationalization came with the Water Act of 1945.

The 1945 Act gave the Minister of Health (now the Minister of Housing and Local Government) three powers related to water withdrawals. First, the Minister was authorized to license new or increased withdrawals of ground water in areas that he might designate as requiring special measures for the conservation of water. Areas so designated were referred to as "conservation areas." Second, the Minister was authorized to require reports on quantity and quality of water taken from any natural water body, whether surface or underground, and in the River Boards Act of 1948 the Boards were directed to collect surface water information for the Minister. Data provided under this authority made available more precise information on amount of withdrawals and need for abstraction control.

A third authority given to the Minister, although only indirectly relevant to abstraction control, is nevertheless significant. This empowers the Minister to create by order a united water district to supply all or parts of districts of one or more local authorities, or to combine water supply undertakings, or transfer responsibilities from one to another. Through this action, major steps have been taken to rationalize service areas and supply areas, including the location and operation of water intake facilities, thus facilitating reporting on, and controlling withdrawals by, chartered water undertakers.

With growing demand for water, particularly for irrigation, new or additional withdrawals threatened other uses in many areas. The Proudman Committee in its final report in 1962 noted with concern that there was in general little or no control over surface water abstractions. In view of increasing withdrawals for industry and irrigation the Committee declared that ". . . it is necessary to apply control to ensure that abstractions do not overstrain the river system at any time and that the benefit of conservation works [reservoirs] is enjoyed by those whose needs they are intended to serve."[35] The Committee recognized abstraction control as a tool of water management, and it recommended that all powers to license withdrawals of water, whether from surface or underground sources, be placed in the new River Authorities.

This recommendation was supported by the government White Paper. In reply to expected allegations of violation of riparian rights to water, the government pointed out that Parliament had already considered it proper to restrict the taking of ground water, and that circumstances of the present day necessitated similar control over the withdrawal of surface water. Nevertheless, it is reported that concern about violating established rights to water generated considerable discussion in Parliament and that the licensing provisions of the Act were subject to many revisions before final enactment.[36]

Administration of withdrawal licenses

The law as it was finally enacted in 1963 imposes a general restriction on water withdrawals, making them subject to a license issued by the appropriate River Authority. Specific exceptions and limitations on the licensing authority are provided in the statute; and action by the Authority is subject to appeal to the Minister.

Exceptions to licensing requirements[37] are provided with re-

[35] Sub-Committee on the Growing Demand for Water, *Final Report* (1962), *op. cit.*, p. 15.

[36] One official reports that although certain common law rights were extinguished by the Act, it was felt at the time that existing rights were sufficiently safeguarded by introducing the concept of MAF, the determination of which must consider the "requirements of existing lawful uses." (Marshall Nixon, Chief Engineer, Trent River Authority, Letter of December 8, 1967.)

[37] *Water Resources Act of 1963*, Secs. 24 and 25.

spect to: (1) withdrawals of less than 1,000 gallons which are not a part of a continuous operation; (2) withdrawals for minor riparian uses such as domestic and livestock supplies and other "on the land" use, *except sprinkler irrigation;* and (3) withdrawals from sources where, upon application for exemptions, it can be demonstrated that licensing control is not needed for that particular supply source.[38]

The latter type of exception is obtained by application of a River Authority, after consultation with the Water Resources Board, to the Minister for an order excepting any one or more sources of supply within the River Authority's area.[39] Such an application presents evidence that licensing requirements of the Act are not needed in relation to the source or sources of supply which are the subject of the application. The Authority submits to the Minister a draft order together with a statement of any observations made by the Water Resources Board. The procedure for approval by the Minister and the role of the Water Resources Board is essentially the same as that required for approving MAF determinations.

The authorization to exempt whole sources of supply reflects quite a different legal-administrative philosophy than that underlying the 1945 Act. In that statute, the Minister was required to designate "conservation areas." He was then authorized to exercise, within those areas, his powers to license ground water abstractions. The procedural requirements for establish-

[38] Other categorical exceptions are made (1) for abstractions required in land and mine drainage; (2) for transfers of water required in the performance of the functions of navigation, harbor, or conservancy authorities; (3) for vessels; (4) for firefighting; and (5) for research and investigations.

[39] The Water Resources Board in its Third Annual Report for the year ending September 1966, cites three such geographic areas for which exceptions were approved in the reporting year. Two pertained to underground sources and one to a surface source in Wales. With respect to the underground sources, the Board reported that "the strata are composed mainly of impermeable rocks and underlie areas where ground-water does not, or is not likely to, make a significant contribution in meeting demands. . . . In each area licensing control has been retained over the more important superficial deposits, commonly sand and gravel, along the river valleys, because abstraction of ground-water from these deposits under particular circumstances can reduce significantly streamflow."

ing "conservation areas" placed the burden of proof for the necessity of abstraction control upon the Minister. Procedural requirements under the 1963 Act in effect place the burden of proof upon the abstractor to show why he should not be subject to abstraction control. In short, the new act assumes the point of view that all withdrawals should be subject to control by the government; then the law proceeds to describe exceptions and procedures for administratively granting exceptions. The shift in philosophy between 1945 and 1963 may reflect a change in the political perception of the seriousness of water supply problems in England.

A more important restriction on the licensing authority in the 1963 Water Resources Act is the recognition of a "license of right." Applicants for a license to abstract water are *entitled* to a license if they qualify in either of two ways: (1) hold a right to abstract water by virtue of any statutory provision in effect prior to the effective date of the 1963 Act, or (2) have abstracted water from the designated source at any time within the period of five years prior to the effective date of the 1963 Act. For applicants meeting either of these two conditions, the River Authorities apparently have no choice but to grant a license, unless established abstractions are found to have been unlawful. The "license of right" considerably restricts the flexibility with which the licensing power can be used to redirect and control water use. However, the licensing process does make mandatory accurate reporting of use, even by those who have a "license of right," and consequently it is expected to effect significant improvements in basic information on water use.

The Water Resources Board points out that abstraction control must be viewed in the context of the general duty imposed by the law upon the River Authorities to take such action as necessary or expedient "for the purpose of conserving, redistributing or otherwise augmenting water resources in their area, or of transferring any such resources to the area of another river authority."[40] The Board recognizes two aspects of the

[40] Water Resources Board, Memorandum No. 1 (WRB-1), "Control of Abstraction and Impounding of Water," Reading Bridge House, Reading, February 23, 1965.

licensing function. First, it declares the objective of abstraction control in the English water management system is to prohibit new abstractions if they would either (a) depress the condition of the inland water to an unreasonable degree, or (b) disturb the "protected rights"[41] of existing abstractors. However, the Board makes clear that abstraction control also has a second purpose, that of protecting and controlling the use of newly developed water supplies.

> There would be no point in conserving and planning the development of water resources if the benefits to be obtained thereby were squandered. Those whose needs give rise to the conservation measures are required by the Act to pay for them. In return, they must be given some guarantee that supplies will be available where they need them, when they need them and in the quantity and quality they require. Moreover existing interests in the river must be protected from over-abstraction or ill-advised impounding. A control over the abstraction and impounding of water is, therefore, fundamental to the proper management of resources for which the Act was designed.[42]

Abstraction licenses may attach such conditions as the River Authority considers appropriate. Specifically, the Act requires licenses to prescribe:

1. Quantity of water authorized, including explanation of the way in which that quantity will be measured and provision for measuring or assessing the amount actually taken during specified period.

2. Location of intakes and for each intake, the amounts, periods of abstraction, means of withdrawal and purposes to be served.

3. The land on which and the purposes for which water abstracted under the license will be used.

4. The period for which the license will be in effect.

---

[41] A protected right is held by anyone abstracting under a license already issued (at this time, primarily a license of right) or under exemptions for domestic and agricultural purposes, for which a license is not necessary.

[42] WRB-1, *op. cit.*

Different provisions may be made by the same license with respect to (a) abstractions during different periods, (b) abstractions at different points or by different means from the same source of supply, and (c) abstraction for different purposes.

The fact that the license may prescribe terminal dates makes abstraction control a more flexible tool of water management than otherwise would be the case. Licenses to withdraw water are, therefore, subject to periodic review, at which times adjustments can be made in conditions attached to the license. The conditions prescribed by law to be attached to a license are applicable to licenses of right as well as to new applicants. Thus, the rigidity of use implied by blanketing in all existing users as a matter of right may gradually be lessened by the periodic review and adjustment procedure.[43]

Although the River Authorities exercise the primary licensing powers in the 1963 Act, their performance is subject to various review and intervention powers of the Minister. For example, the Minister can intervene and take action instead of an Authority. He may do this by directing that applications for licenses (except licenses of right) be submitted to him instead of to a River Authority. Such a direction may apply to a specific River Authority or to all Authorities; it may apply to a particular application or to a class of applications. This is a summary power of considerable significance.

In addition, specific provisions grant rights of appeal to the Minister from actions or inactions of River Authorities. The Minister is authorized, and in some cases required, to conduct local inquiries and/or hold hearings as a basis for his actions. The action of reviewing, revoking, or modifying a license is subject to essentially the same procedures as approval of the original license, including authority of the Minister to direct the review, revocation, or modification of licenses.

---

[43] There appear to be some differences in interpretation as to whether a license of right can be terminated or modified through the revision process herein described. An official of one River Authority believes a license of right can be changed only under revocation procedures which require full compensation for any financial equity reasonably attached to the right. (Marshall Nixon, Chief Engineer, Trent River Authority, Letter of December 8, 1967.)

Levying Abstraction Charges

The Water Resources Act of 1963 gave the River Authorities in England and Wales extensive new powers to levy charges on water withdrawals. Prior to the Act the only assessments authorized which had any direct relationship to users benefits were those on internal drainage boards and fees attached to the issuance of fishing licenses. Other activities of the former River Boards were financed by precepts (a millage assessment) on appropriate counties and spread upon their taxable property.

In addition to these inherited powers of assessment, the 1963 Act authorizes River Authorities to levy charges on water withdrawals. Each Authority is required to prepare a "charging scheme" based upon the quantity of water licensed for abstraction.[44] Charging schemes are approved by the Minister for a period of five years. The law authorizes differential rates based upon four "relevant circumstances": (1) the characteristics of the source of supply from which the water is authorized to be abstracted, (2) the season of the year at which the water is authorized to be withdrawn, (3) the purposes for which, in accordance with the provisions of the license, the water is authorized to be used, and (4) the way in which the water is to be disposed of after being used for the purpose specified in the license. Rates approved in "charging schemes" are maximum rates for the five-year period for which each scheme is approved. Charges actually levied under the scheme in any financial year are to be at such rates (not exceeding the maximum specified in the scheme) as are necessary to meet costs that are chargeable to this source of revenue. Such costs are principally those associated with constructing and operating conservation impoundments.

In practice then, a public or private water undertaker acquiring a license would pay the River Authority a "wholesale" use charge based upon the quantity of water he was authorized to withdraw. Presumably this charge would be passed on to retail

[44] A subsequent section of the Act provides a procedure whereby irrigators may be granted the privilege of paying for water used in spray irrigation on the basis of quantity actually used, instead of the quantity licensed (Sec. 63).

consumers either by meter charge or a levy on the ratepayer (property taxpayer) of the service area, thus adding to the present "production cost" of water services. Irrigators and industries holding abstraction licenses may, as ultimate users, pay their water bill directly to the River Authority.

River Authorities are also authorized to "make such reasonable charges as they may determine in respect of the use of any reservoir (under management of the Authority) for the purposes of recreation, and of any facilities made available by the Authority." This authorization has been little discussed, perhaps because the possession of management control of reservoirs (either by construction or agreement with other owners) by River Authorities was still in the future at the time this study was underway.

Administration of abstraction charges

The authorization for levying abstraction charges also prescribed administrative procedures relating to establishing rate structures, determining annual levy, and granting exemptions.

As soon as practical after the effective date of the Act, each River Authority is required to prepare a charging scheme which sets forth a schedule of maximum rates the Authority proposes to levy for each category of user, based upon the relevant circumstances set forth in the Act. The law requires each River Authority to have its charging scheme approved by the Minister on or before April 1, 1969, five years after the effective date of the Act. Until an official charging scheme is adopted, licensed abstractors (except those holding a license of right) may be required to pay the River Authority an abstraction charge, which, if it cannot be agreed upon between the Authority and the abstractor, will be set by the Minister.

Procedure for approval of a charging scheme follows that required in determining minimum acceptable flow.[45] A draft statement setting forth the proposed charging scheme is prepared by the River Authority. It is made available for review by the general public for a twenty-eight-day period. Comments and

[45] Procedure for setting minimum acceptable flows is described earlier in this chapter in the section on "Administration of MAF."

criticism of the proposal by members of the public may be directed to the Minister. The River Authority is required to give notice and to make available upon demand copies of the charging scheme as proposed to the Minister. Objections received may cause the Minister to conduct an inquiry and/or hold hearings. The Minister may approve the scheme as proposed or may alter it in such manner as he decides. Such alterations are subject to the same notice and hearing procedures as apply to the original proposal.

Revisions of charging schemes may be initiated after five years of operation of a duly approved scheme by either the relevant River Authority, a statutory water undertaker subject to the scheme, or a group of ten or more licensed abstractors within the River Authority area. Revisions are subject essentially to the same procedures for approval as apply to the original proposal.

A River Authority may, upon application of an abstractor, exempt applicant from payment or agree to a reduced charge, based upon the applicant's material contribution to the conservation responsibilities of the River Authority. Typically, such a privilege might be exercised by a statutory water undertaker who owned and operated an impoundment in the River Authority area, if the reservoir was or could be incorporated into the Authority's scheme for the augmentation of water supplies for the area. The Minister is empowered to *direct* the River Authorities in this regard and to hear and act upon appeals either by the applicant or by the Authority.

Actual charges to be levied each year under an approved charging scheme are determined by the River Authority as an amount necessary, "taking one year with another," to meet expenses chargeable to the Authority's water conservation responsibilities. Each River Authority is required to keep a separate water resource account for the capital financing of water conservation measures. Annual levies are based upon amounts necessary to keep the water resource account in balance. Accounting methods and operations are audited by the Minister as a part of his general responsibilities over the operations of local governments.

Construction and Co-ordination of Water Development

In general, central government's role in the developmental aspects of water resources in England has been late and timid. The use of reservoirs to regulate rivers for multiple benefits has had comparatively little attention until recently. A dam in England has more often been constructed and operated to regulate water levels for navigation or to provide a reservoir for water supply purposes. Water delivery for supply purposes is often by conduit and not infrequently to another drainage. The primary constraint on the construction and operation of storage reservoirs has been the requirement for the release of "compensation water," theoretically in amounts necessary to satisfy riparian rights down stream.

The Proudman Committee in 1962 expressed considerable concern about the unco-ordinated patchwork of single purpose reservoirs, and it recommended that the River Authorities have legal power and responsibility to develop and operate systems of water control structures to augment the use capacity of hydrologic units in their respective areas. Accordingly, the Water Resources Act contains several new provisions which empower the Authorities to construct and co-ordinate operations of water development and management facilities.

Construction

Construction of water developments, particularly reservoirs, has largely been a matter for private enterprise and local public authorities. There are three exceptions where central government has assumed developmental responsibilities. These are the early interventions in navigation improvements, in land drainage (including shore protection)[46] and, since 1923, in the provision of facilities to improve fisheries.[47] Responsibility for public water supplies was early given to private water undertakers.

[46] See Chapter 3 for fuller description of the early involvement of government in navigation and land drainage.

[47] The Salmon and Fresh Water Fisheries Act of 1923 established Fishery Boards in designated river basin areas, authorizing them to collect license fees and purchase or lease any dam, weir, or any other fishing facility, and to construct and maintain fish passes in dams or abolish or modify existing passes.

In the more recent past, local authorities have sometimes taken over operations as public water undertakers. Water undertakers were brought under a degree of government supervision by the Water Act of 1945. Under that statute, the Minister of Housing and Local Government has far-reaching authority to influence water undertakers, private and public, including authority to advise on their planning of future developments, to reconcile their sources of supply and service areas, and to require them to supply water for non-domestic uses within their service area.

The wording of the 1945 statute gives the impression that great care was exercised to avoid any charge of unjustified infringement on the "rights" of water undertakers to construct and operate water supply reservoirs. Even the Milne Committee, which in 1943 recommended the creation of River Boards as a response to postwar needs, specifically recommended against the Boards being empowered to construct reservoirs for the conservation of water. The Committee stated ". . . apart from the question of the considerable expense involved, the exercise of such a power might conflict with the rights of riparian owners or the statutory powers of other Authorities. We consider, therefore, that if such works are found to be required in any area, the Board concerned should promote local legislation for the purpose."[48] The concept of individual local responsibility for development was publicly articulated by this Committee at the same time that it advocated strong integrated river management responsibilities for the River Boards.

In contrast, the Water Resources Act of 1963 gives to the new River Authorities comprehensive powers for constructing, operating, and financing development projects. It gives the local river agencies all the construction powers previously held by River Boards (including means of financing), relating primarily to land drainage, flood control, and fisheries (and navigation in special instances). In addition, the River Authorities are newly empowered "to carry out such engineering or building operations as they consider necessary or expedient for the purposes of any of their functions" (Sec. 69), and "to acquire by agreement

---

[48] Central Advisory Water Committee (Milne Committee), *Final Report* (1943), *op. cit.*

any land which they required for any purpose" related to their functions under the Act, including acquisition by condemnation, which shall have approval of the Minister (Sec. 65). Although the new provisions in the Act appear to provide a blanket construction authority, it is qualified by many other procedural requirements such as (1) the requirement to obtain "planning permission" from the relevant local or regional planning agent, (2) the necessity to obtain Ministerial consent for land condemnations, (3) the need to obtain approval to borrow money, and (4) the fact that the River Authorities actually can construct projects only for those purposes for which they are authorized to finance.

A new system of financing river developments is provided by the introduction of abstraction charges under authority of the 1963 Water Resources Act.[49] The income from abstraction charges goes into a special water resources account, expenditures from which may be made to meet that portion of costs of water development facilities chargeable to water conservation. This source of funds for capital investments is, of course, supplemented by the power (transferred from the earlier River Boards) to precept local taxing authorities as a means of financing the local share of drainage, flood control, and general administration costs. Grants from central government are available for drainage and flood control. (For a graphic representation of the general relationships involved in the aggregate system of financing, see Figure 5, Chapter 5.)

It seems clear that government in England has been slower to move into direct construction and operation of river basin development than it has to extend governmental regulatory powers to influence water resources use. The 1963 Water Resources Act was, in the main, a victory for those who believed the time was overdue for the government to take and emphasize a positive management approach in contrast to its longtime emphasis upon regulation.[50] In the context of the English ex-

---

[49] Section 58.

[50] This was one of the major themes of the Proudman Committee. See *Final Report* of the Sub-Committee on the Growing Demand for Water, *op. cit.*

perience, the 1963 legislation must be credited with providing River Authorities extensive and comprehensive powers for the construction and operation of river basin developments.

Co-ordination

In addition to their responsibilities for direct construction and operation of projects, the River Authorities were given several related powers which provide them with tools to secure co-ordination of projects not under their ownership but which may affect the flow and availability of water. Four specific powers are significant in this regard:

1. Construction or alteration of any impoundment in any inland water in a River Authority's area is prohibited except under license granted by the Authority. The license may prescribe the amount and manner in which the impoundment will impede the flow of water (Sec. 36). This restriction does not apply to construction by navigation authorities, harbour authorities, or conservation authorities. Presumably, it will operate primarily with respect to new developments by water undertakers and industry.

The River Authority's power to license impoundments is qualified by two responsibilities of the Minister. First, the Minister may by order take over licensing with respect to two categories of applications: (1) for a particular River Authority or for River Authorities generally, and/or (2) for particular applications or for a class of application (Sec. 38). Second, licensing actions taken by River Authorities are subject to appeal to the Minister.

2. A River Authority may enter into agreement with statutory water undertakers, local authorities or land occupiers with respect to construction and maintenance of works and the manner in which any reservoir in the River Authority's area will be operated. The Minister may direct that agreements, in general or in particular, have his approval and/or have been cleared with the Water Resources Board (Sec. 81).

3. A River Authority may enter into an agreement with a navigation authority, harbour authority, or a conservancy au-

thority for interagency exchange of payments in accordance with benefits received by one from project works of another. In case the two authorities cannot agree to payments appropriate to the benefit-cost exchange between their operations, the matter is referred to the Minister, and to the Minister of Transport in case of harbours and navigation. Their decision is final, and respective authorities are obliged to pay accordingly (Sec. 91).

4. A River Authority is authorized to request the Minister to make an order transferring to the River Authority any of the following:

(a) functions and property of a navigation authority, harbour authority or conservancy authority.

(b) reservoirs, wells and boreholes belonging to statutory water undertakers if not operated primarily for water supply.

(c) the functions of managing and operating reservoirs, wells and boreholes as provided in (b), but without a transfer of ownership of the reservoir or the work.

This authority becomes reciprocal whenever a River Authority initiates action under it, that is, when a River Authority applies for an order transferring any functions or property of another body, "that body may themselves apply for such an order" (Sec. 82).

The combination of functions relating to water development projects gives to the River Authorities the legal power to wield a strong influence and control over construction and operation of projects in their respective areas, whether under their ownership or under private or other public organizations. Through these powers for construction and co-ordination, a River Authority has the potential of integrating the planning, implementation, and operation of works designed to control and make available the water resources under its jurisdiction.

5

ORGANIZING AND FINANCING WATER MANAGEMENT IN ENGLAND

As a charter for water management, the Water Resources Act of 1963 is as important for its organizational and financial provisions as for its new legal powers. Both organizational arrangements and the provisions for financing are instrumental to the substantive water management measures discussed in Chapter 4.

Organizational Arrangements

Three central agencies and twenty-nine River Authorities are the principal units in the water management system. The Ministry of Agriculture, Food and Fisheries, and the Ministry of Housing and Local Government have long been the primary cabinet agencies in water resources affairs in England. The Water Resources Board was created by the Water Resources Act to play a major leadership role in water conservation. The River Authorities are the fourth and key element of the institutional system, and their role is central to the description and analysis which follows.

River Authorities

The River Authorities have a general mandate to take such action as necessary or expedient, or as they may be directed, "for the purpose of conserving, redistributing, or otherwise augmenting water resources in their area, or of transferring any such resources to the area of another river authority."[1] In most cases the River Authorities are the initiators of water manage-

[1] *Water Resources Act of 1963*, Sec. 4.

ment actions and have a wide latitude in decision making. Within its respective area, each River Authority is authorized to:

1. Provide for systematic collection of hydrologic data and of water use information.

2. Conduct water resource surveys, including analysis of supply and demand relationships, and development of proposals for actions.

3. Determine a "minimum acceptable flow" (or level) at critical points in the water systems of their areas to serve as a standard to guide decisions regarding water management.

4. Administer pollution control legislation, including particularly the issuance of "consents" to discharge waste water into the natural water bodies.

5. License withdrawals from ground and surface water resources.

6. Maintain surveillance of fishing conditions and issue fishing licenses.

7. Plan, construct, and operate water developments for the purpose of land drainage, flood control, fisheries improvements, water conservation for domestic, industrial, and agricultural purposes, and, in some instances, for navigation and harbour improvements.

8. Co-ordinate non-Authority water development projects to better achieve their contribution to an integrated scheme of water management.

9. Secure funds to finance water resource developments and to cover costs of their operations and administrative needs of the River Authority itself.

A River Authority's powers of water management and the manner in which central government supervises the performance of these functions were elaborated in Chapter 4.

A River Authority is composed of from twenty-one to thirty-one members, made up of representatives of local governments and appointees of appropriate Ministers of central government chosen for their qualifications in drainage and flood control, fisheries, agriculture, public water supply, and industry. In a sense, the River Authority is an instrument for creating a partnership between local governments and the national government in designated water regions. The law provides that appointees of local government and of central government will be in such proportions as always to assure local governments a bare majority, but no more. Whom the members of River Authorities represent is central to the question of their effectiveness as regional water management agencies. The representation issue is considered more fully in a final section of this chapter.

The Water Resources Board

The newly created Water Resources Board provides for the first time a central focus for the national concern about water resources and their effective development and management. The Board is composed of eight members who are appointed by the Minister of Housing and Local Government. A chairman and deputy chairman are designated by the Minister. The Board may appoint such staff as the Minister, with Treasury consent, approves. The law requires that the Board have at least one member "having special knowledge and experience of matters relating to the conservation and use of water resources in Wales." No other qualifications for Board membership are prescribed. The Water Resources Board, organizationally, is in the Ministry of Housing and Local Government, but it is essentially independent in exercising its statutory authorities.

In general, the powers and functions of this new central water resources agency fall into two categories: (a) formulating and expressing a national perspective regarding water resources, and (b) giving central leadership to the River Authorities in performing their "new functions," that is, those functions relating to water conservation.

In executing its responsibility to provide a national perspective, the Board is authorized to perform the following functions:

1. Collect, collate, and publish data and information relating to water resources and prospective demands upon the resources.
2. Advise the Ministers regarding their joint and separate re-

sponsibilities "to promote the conservation and proper use of water resources and the provision of water supplies . . . and to secure the effective execution by water undertakers . . . of a national policy relating to water . . ."[2]

3. Recommend action necessary for conserving, redistributing, augmenting, or securing the proper use of water resources.

4. Carry out or sponsor such research, make such inquiries, and submit such reports as the Board considers necessary, or as a Minister may request.

In providing central leadership to River Authorities, the following general responsibilities are given to the Water Resources Board:

1. Upon request of a River Authority, advise with respect to the Authority's water conservation responsibilities,[3] including, for example, the collection of hydrologic data; the conduct of water resource surveys; the formulation of plans and proposals for action; the determination of minimum acceptable flows; and the licensing of water withdrawals and reservoir construction.

2. Review and recommend to a River Authority and/or the Minister revisions in actions taken by an Authority relating to its water conservation responsibilities.

3. Encourage and assist River Authorities in the development of plans to transfer water from one area to another, and approve proposals for such transfers.

4. Review progress under the Prevention of Pollution Acts of 1951 and 1961, and recommend action where, in the Board's judgment, specific waters need improvement which could be realized through the exercise of a River Authority's powers under these Acts to regulate effluent discharges.

5. Report to the Ministers cases where, in the Board's judgment, River Authorities are in default of their responsibilities;

[2] *Water Resources Act of 1963,* Sec. 12, and Sec. 1.

[3] The reader should be reminded that "conservation responsibilities" in this context refers to responsibility for the adequacy of water for withdrawal uses, primarily for municipal, industrial, and irrigation purposes. The Water Resources Board has little involvement in land drainage, flood control, fisheries, or administration of consents for effluent discharges. These functions were transferred from the earlier River Boards.

and if a defaulting authority fails to comply with a Ministerial order, the functions of the River Authority relating to water conservation may be transferred to and exercised by the Board. (Other functions of defaulting River Authorities may be transferred to the appropriate Ministers.)[4]

In summary, it may be said that the Water Resources Board has responsibility for a general surveillance of River Authority actions relating to water availability, water quality control, and the development and implementation of water conservation schemes. The Minister of Agriculture, Food and Fisheries is still responsible for supervising matters relating to drainage, flood control, and fisheries, and the Minister of Housing and Local Government continues to exercise supervision of the River Authorities in their dealings with water undertakers and local governments on matters of water supply services and in their administration of waste water discharge permits.

Although the Water Resources Board shares policy and program leadership with the two Ministries, the Board's influence may well extend far beyond the specific functions noted in the Act. The Board, chosen to be broadly representative of the whole scope of water affairs, has set out to assemble a professionally competent staff through which to perform its functions.[5] With the dominance of concern for water conservation including water quality improvement, which falls clearly in the Board's purview, this new agency may quite quickly become the central agency on most questions of water management.

The Board has assumed a positive leadership role in providing River Authorities with policy and procedural guides. In its first full year of operation, it developed six comprehensive memoranda clarifying, interpreting, and suggesting procedures for carrying out the new statutory provisions. Some of these memoranda have been the occasion of two- and three-day conferences with appropriate officials and/or employees of the

[4] *Water Resources Act of 1963*, Sec. 108.

[5] By September 30, 1966, the Board had 97 authorized positions, 42 of which are in the administrative and clerical categories. *Third Annual Report of the Water Resources Board* (Appendix A).

River Authorities.[6] These first-year memoranda covered the following subjects: (1) control of abstraction and impounding of water; (2) periodical surveys of water resources, including twenty-year demand projections, and proposals for necessary action; (3) hydrometric schemes; (4) water in underground strata; (5) hydrometric schemes related to water quality; (6) control of abstractions of water in underground strata.

The Cabinet Ministries

Two Ministries in the central government have both general and specific responsibilities regarding water resources.

The Minister of Housing and Local Government exercises the central government responsibilities for municipal water supply and sewage disposal. Of prime importance are the Minister's powers relating to (1) the rationalization and operation of water supply services provided by private water undertakers and local governmental units, and (2) the provision of sewerage and sewage treatment systems by local authorities. In addition, the Ministry exercises important supervisory functions regarding certain operations of River Authorities. Four major areas of Ministerial supervision are the determination of "minimum acceptable flows," the licensing of abstractions, the approval of charging schemes, and the granting of consents for effluent discharges. These powers regarding water use and development are associated in the Ministry with a wide range of supervisory authorities over the operations of local governments, including those relating to finance and planning.

The Minister of Agriculture, Fisheries and Food has general and specific authorities over River Authorities relating to drainage, flood control, and fisheries. Important among these is the responsibility to recommend grants-in-aid for drainage and flood protection projects, and to provide policy guidance in the administration of fishing regulations and fishery improvements on inland waters within River Authorities' areas of jurisdiction.

The two Ministries share responsibilities for general supervision of the River Authorities, particularly with respect to their establishment, membership, and general operating policies.

[6] *Second Annual Report of the Water Resources Board*, September 1965.

In general, the Ministries are empowered to exercise three kinds of administrative constraints over River Authority operations. The River Authorities cannot act on certain proposals until they obtain Ministry approval. Appeals against actions taken by River Authorities come before the Ministries for hearing and arbitration. Also, the Ministries can override certain actions of the River Authorities by exercising either special powers or their general default powers. General default powers[7] provide that if an inquiry shows that a River Authority has failed to perform any of its functions in a case where it should have done, and if it does not act after it has been requested to do so, the Ministers may transfer such functions as they deem appropriate to a Minister, the Water Resources Board, or an adjoining River Authority.

In fact, Ministerial powers regarding water management are basically quite limited. A minimal number of approvals are required, and only on key program actions. Ministerial action is taken on appeals from River Authority actions, and the Ministers may decide to intervene under permissive powers in specific kinds of actions. But the Ministers have little authority to initiate action except under default powers. (The relationships between the Ministers, the Water Resources Board and the River Authorities in administering the principal governmental powers for water management were developed in greater detail in Chapter 4.)

Financing Water Management

The provision of legal and administrative measures for financing is one of the major aspects of institutional arrangements for water management. Financing measures may be expected to serve two not necessarily consistent purposes. First, they are viewed by a water management agency as a source revenue to support its activities; and second, they may be considered a device for beneficiaries to pay some part of the value they derive from water services enjoyed. This distinction is not usually made either in designing or in evaluating financing methods,

[7] *Water Resources Act of 1963*, Sec. 108.

and in practice each measure may, to greater or lesser extent, serve both purposes. In this chapter, the concern is primarily with sources of revenue for River Authority operation. The question of payments by beneficiaries as an incentive to efficient use of water services is considered in Chapter 6 as a means of regulating water use.

The Water Resources Act of 1963 provides two general categories of revenue for financing public water management efforts. The first is the group of financial provisions associated with earlier functions which had been exercised by River Boards and are now transferred to River Authorities. The second is a new legal authority to institute a system of charges upon water withdrawals.

Background

Four types of financial provisions applicable to water development activities existed prior to the Water Resources Act of 1963. These evolved from the gradual involvement of government in water resource affairs in England through the last two centuries. Because these inherited financing measures are important to understanding the institutional relationships involved in water management under the Water Resources Act of 1963, they are summarized in the following paragraphs:

1. Statutory Water Undertakers have been authorized for more than a century to construct and operate water supply systems and to levy and collect water rates[8] from those receiving the service. The Water Act of 1945 made no basic change in this aspect of financing. However, it did update methods of regulating the financial operations of the undertakers.[9]

Specifically, the act prescribes maximum charges, and authorizes the Minister to alter rate schedules and give relief from the ceiling; it authorizes the Minister to approve capital shares issued by companies; and it prescribes restrictions on company dividends.

---

[8] Water rates are, in effect, assessments based upon a proportion of property valuation rather than upon amount of water withdrawn or other index of value received.

[9] *Water Act of 1945*, Sec. 40; schedule 3, parts XII and XV.

2. From early times, land drainage has been financed on the principle of "no benefit—no charge." Charges were levied as a rate by internal drainage boards[10] on those properties directly benefited by drainage improvements. In 1930, Catchment Boards were established to cover larger drainage systems within which might operate several internal drainage boards. The Catchment Boards were authorized to construct and operate larger improvement works on the main river channel and to precept[11] internal drainage boards for benefits enjoyed from improvements on the main river undertaken by the Catchment Boards. This authority was later transferred to the River Boards, including the authority to precept County Councils and County Borough Councils to cover the broader indirect benefits from drainage, including flood control in urban centers. Precepts on Counties and County Boroughs often amounted to as much as two-thirds to three-fourths of the revenue of a River Board.

In 1961, River Boards' financing powers relating to drainage and flood control were expanded. At this time they were authorized to levy charges directly on agricultural lands for two purposes: (1) to assure that those lands contribute to general benefits of Main River works since farm lands do not pay general rates to local rating authorities and thus do not contribute through precepts on the County Councils (farm lands were "derated" during the depression); and (2) to finance agricultural drainage works on water courses not designated as Main River nor included in an Internal Drainage District. Thus, the River Boards were in effect given power to levy drainage rates directly on the property owners or occupiers as the primary beneficiaries.

3. Fishery Boards created under the Salmon and Freshwater Fisheries Act of 1923 were authorized to issue fishing licenses and collect license fees for fishing in their respective areas. With these revenues the Fishery Boards were authorized to acquire, construct, and maintain capital facilities for fishery purposes. Those development and financial powers were transferred to River Boards in 1948.

[10] Organizations of local property owners whose lands are potentially affected by land drainage schemes for the purpose of levying drainage rates.
[11] See "Definitions," Chapter 1.

4. Various harbour boards, navigation authorities, and conservancy authorities have been established by Royal Charter and by Acts of Parliament to develop and maintain navigational waterways. In principle, these have been financed by user charges. The River Boards were authorized to assume the functions of those authorities where appropriate and under specified circumstances.

Present sources of revenue

The scope of powers available to River Authorities for financing water management activities may be classified as follows:

1. Precepts, that is, assessments levied against other legally recognized local authorities and special districts.

(a) Precepts on Internal Drainage Boards for benefits derived from works on the Main River, undertaken by the River Authority. (The Drainage Board in turn raises its funds from special drainage rates levied against those lands directly benefited.)

(b) Precepts upon local authorities, specifically County Councils and County Borough Councils, for three purposes:

(i) general benefits of drainage and flood control derived from works of the River Authority on the Main River.[12]

(ii) temporary capital financing of water conservation works, that is, storage and regulation related to the water supply function. Authorized by Water Resources Act of 1963 until 1969 when charging schemes (see 2(b) below) based on quantity of water abstracted are to be in effect.[13]

(iii) general expenditures not otherwise specifically provided for, including costs of administering pollution prevention acts of 1951 and 1961.[14]

2. Collection of fees for water use.

(a) Fishing license fees. River Authorities, by virtue of powers under the 1923 Salmon and Freshwater Fisheries

[12] *Land Drainage Act of 1961.*
[13] *Water Resources Act of 1963*, Sec. 87(1)b.
[14] *Water Resources Act of 1963*, Sec. 87(1)a.

Act, are authorized to charge £1 per license or such fees as the Minister of Agriculture may approve. Fishing license fees have been a minor item of total income of River Board operations. In all except one River Board, fishing expenditures tended to exceed license income by at least a small amount.

(b) Charges for water withdrawals. The Water Resources Act of 1963 directs the River Authority to institute a charging scheme (by April 1, 1969) based upon the quantity of water a licensee is authorized to take.[15]

(1) Differential rates are authorized, based upon four relevant circumstances (i) characteristics of the source, (ii) season of the year, (iii) proposed use of the water, (iv) methods of disposal.

(2) The River Authority, with the Minister's approval, may make individual exceptions from full charges on the basis of benefits from works constructed or to be constructed (or other financial contributions) by licensees (Sec. 60).

(3) New abstractors will be charged for abstractions immediately upon licensing even though comprehensive charging schemes have not yet been put into effect (Sec. 62).

(c) Harbour and navigation fees when a River Authority exercises the legal authorization to assume responsibilities of a harbour board, a navigation authority, or a conservancy authority. Only two River Boards exercised their powers in this regard. This is likely to be a minor source of funds except in the case of the Thames Conservancy, which, although not a River Authority, does exercise essentially the same powers and perform the same functions as if it were a River Authority.[16]

3. Assessment, on the basis of rated value, of land owners (a) for *direct benefit* from drainage work on water courses not classified as Main Rivers nor in the jurisdiction of an Internal Drain-

---

[15] *Water Resources Act of 1963*, Sec. 58.
[16] *Water Resources Act of 1963*, Sec. 82(1)a.

age Board and (b) to get *general benefit* contribution to costs of Main River works from agriculture lands which are exempt from general rates and cannot be tapped through County Council and Borough Council precepts described above.[17]

4. Exchange payments among authorities. Navigation, harbour, or conservancy authorities may make benefit payments to the River Authorities as may be agreed represent benefits derived from works and operations of the River Authority. Reciprocally, a River Authority may make payments to other authorities for benefits received or services rendered.[18] In practice, River Boards have also negotiated exchange payments with local (municipal) authorities and internal drainage boards, and there is no reason River Authorities cannot continue this practice under general powers.

5. Grants from central government.

(a) Land drainage and flood control

(1) A River Authority may receive grant-in-aid for some fraction of the costs for Main River works. Amount of the grant depends upon the capacity to pay of the counties and county boroughs on whom the remaining costs will be precepted. In practice such grants vary from 0 to 90 per cent of total costs. It was estimated that these grants average 60 per cent of total costs and that total costs of schemes that would qualify for grant assistance under this category for the 10-year period, 1953–63, was approximately £6,582,000.[19]

(2) A River Authority may receive grant-in-aid of 50 per cent of project costs on all works for which there are direct agricultural drainage benefits (in the same manner as an internal drainage board). In general, River Authorities would qualify for this grant only for those works which qualify for special drainage assessment under the Land Drainage Act of 1961. (See 3-b, above.)

---

[17] *Land Drainage Act of 1961.*
[18] *Water Resources Act of 1963*, Sec. 91.
[19] Interview, Ministry of Agriculture, Food and Fisheries.

(b) River basin hydrometric schemes

The Minister of Housing and Local Government may certify grant-aid to River Authorities to help in financing the hydrometric scheme which each Authority is required to carry out.[20]

(c) Hydrologic research

Grants may be made to River Authorities by the Water Resources Board, with approval of the Minister, to cover the costs of (1) obtaining hydrologic data requested by the Water Resources Board in addition to those data considered a part of a hydrometric scheme, and (2) conducting such experimental work in hydrology as may be requested by the Water Resources Board.[21]

6. Borrowing. River Authorities may borrow temporarily by overdraft or otherwise; or negotiate long-term mortgages in order to (i) acquire land, (ii) construct works, (iii) repay money previously borrowed by River Boards or (iv) pay for other things, the cost of which should be spread over a term of years. Decisions to borrow are usually subject to the approval of the appropriate Minister. Borrowing powers of River Authorities are subject to appropriate control provisions of the Local Government Act of 1933, which generally regulates financing power of counties, county boroughs, and municipalities.[22]

Two types of grants to local authorities may have significant impact upon water management within a River Authority area. The first type includes grants for water supply and waste disposal facilities. Grants can be made for these purposes specifically under two statutes:[23]

1. A Rural Water Supply and Sewerage Act provides grants designed to compensate rural areas for those increments of cost in community water supply and sewerage stemming from low density and/or isolation. Grants do not, therefore, encourage treatment plants or other facilities in rural areas that are not at a cost disadvantage.

---

[20] *Water Resources Act of 1963*, Sec. 89.
[21] *Water Resources Act of 1963*, Sec. 90.
[22] *Water Resources Act of 1963*, Sec. 92.
[23] Interview, Ministry of Housing and Local Government.

2. A local Employment Act provides grants to facilitate economic development in local areas through improving the amenities of a locality whose employment rate is below the national average. In some circumstances, water supply systems or sewage treatment systems may qualify.

It seems clear that grants-in-aid specifically for water supply and waste disposal are likely to be quite restricted in their effects upon water management in general.

The second type of central government grant may be of greater significance to water management. Included in this type are two, widely used, general-purpose subventions.

1. Block grants, instituted in 1958 to replace a growing variety of special-purpose grants, are made each year in lump sum to individual local authorities as an aid to local government. Each grant is adjusted annually on the basis of a "budget request" specifying needed amounts by purposes. To what extent investment for local water supply and sewage disposal have been or may be aided by these grants is not readily determined.

2. Rate-deficiency grants may have an extensive influence on water resource developments. The purpose of this system of grants is to even out the inequalities in rateables (that is, local property tax resources) among local taxing authorities. Pursuant to the Local Government Act of 1958, Counties and County Boroughs are assigned an index based upon rateable values per capita. Those local authorities whose index is below the average for the nation qualify for rate deficiency grants.

Rate-deficiency grants would appear to have two consequences for water development investment by local authorities: (i) River Authority precepts on local authorities are based either on actual rateable values or standard (average) rateable values, whichever is higher. Below average local authorities are precepted on the basis of the average, and the difference may be received by the local authority in its rate deficiency grant.[24] This differ-

[24] This is based upon the author's interpretation of Section 93(3) of the Act which reads, "There shall be defrayed out of moneys provided by Parliament any increases attributable to the Act in the sums payable out of moneys so provided by way of Rate-deficiency Grant . . . under the enactments relating to local government. . . ."

ence amounts to a central government subsidy to River Authority operations through the channels of rate-deficiency grants. (ii) At the same time each local authority whose index is below average is, in addition, qualified for a general rate-deficiency grant, which may, at the local authority's discretion, be used for water supply or sewage disposal facilities.

In the years ahead, the combined operations of River Authorities will, in general, be supported by funds from three primary sources: (1) precepts on local authorities for a variety of purposes, (2) water user charges levied on licensed withdrawals, and (3) central government grants. Figure 5 depicts the general relationship among these three types of revenue sources for River Authorities.

It is notable that, except for drainage and hydrometric schemes, the major subventions are to local jurisdictions of general government rather than to the River Authority. Any influence on regional water management schemes is therefore effected through local governments which also collectively maintain a majority control in the River Authority.

Representation on the River Authority

A striking feature of the institutional system for water management in England is the extensive delegation to the River Authorities of water management responsibility. Attention therefore is now directed to the source of power for River Authority decisions.

The law places primary responsibility for initiating regional water management actions in the River Authority—a collective of individuals, appointed as prescribed by the Act, to constitute a "body corporate with perpetual succession and a common seal." Considerable importance therefore is attached to who serves on the Authority, what interest each member represents, and how he participates in the decisions. Indeed, the composition of the River Authority was the major point of issue in the evolution of the Water Resources Act of 1963. The general nature of the issue is reflected in the final report of the Sub-Committee on the Growing Demand for Water, in the minority report attached thereto, and in the Government's White Paper

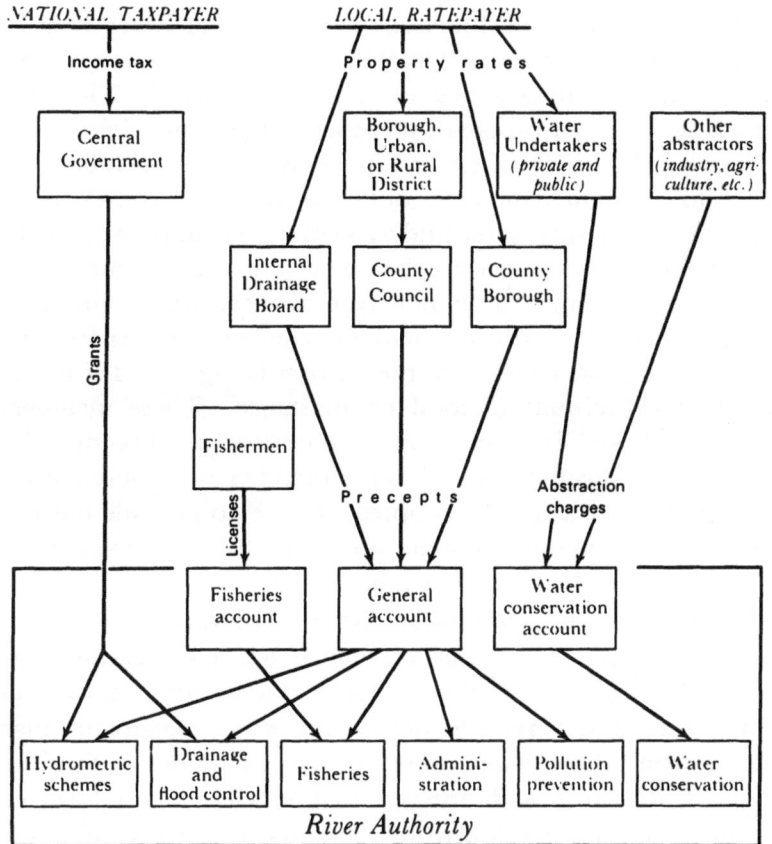

FIGURE 5. METHODS OF FINANCING RIVER AUTHORITIES IN ENGLAND AND WALES.

on the sub-committee's recommendations. Relevant highlights from these reports are reviewed, and the representation question is discussed in the remainder of this chapter.

The representation issue in the sub-committee

Central to the sub-committee's recommendations on water policy was the belief that the time had come for the "creation of comprehensive new authorities to manage the water resources of river basins as a whole and to be charged with a positive duty of water conservation."[25] It was recommended that the River

[25] Central Advisory Water Committee, Sub-Committee on the Growing Demand for Water, *Final Report* (HMSO, 1962), p. 2.

Authorities assume the functions and powers of River Boards, in addition to the new responsibilities regarding water conservation which were specified in considerable detail. The sub-committee rejected proposals merely to enlarge membership of River Boards, which often had as many as forty members, in order to accommodate representatives of water conservation interests, principally water undertakers (water suppliers), industry, and irrigators. Instead, the sub-committee recommended "the creation of an entirely new type of authority," saying that the new situation "calls for small and compact authorities with a membership of, say, 10–15, the pattern being varied if necessary to take account of local circumstances. These members could be elected by the interests concerned or appointed by Ministers in consultation with those interests, or a combination of these methods could be adopted, and we do not rule out the possibility of appointing some members able to devote their full time to the business of the authority."[26]

It is possible that the sub-committee's inspiration for "an entirely new type of Authority" came from a delegation that visited West Germany and was impressed with the water associations prominent there. The delegation, while recognizing that the German arrangements were not necessarily the pattern for England, did note with interest that in Germany a form of organization "has developed for the authorities responsible for conservation measures which includes a large representative body covering the various interests concerned (industry, municipalities, landowners, and others) together with a small executive committee which is responsible for day-to-day administration."[27]

The sub-committee position regarding the composition of river authorities stresses the importance of a small governing board, holding that any appropriate addition to river board membership was "bound to result in bodies too cumbrous for efficient working." Accordingly, the sub-committee seemed to give "representation" secondary consideration. The report in fact states, "We attach less importance to representation as such

[26] *Ibid.*, p. 24.
[27] *Ibid.*, p. 42.

than to efficient and expeditious discharge of the wide range of functions involved." In elaborating its position the sub-committee goes on to say:

> We realise that our recommendation involves a departure from a principle underlying the present constitution of river boards, that representation is partly related to the provision of finance. We think this is unavoidable. The present functions of river boards are financed largely by precepts on rating authorities. Conservation will be financed by charges or licence fees levied on a wide range of new interests, the water users, who will expect fair consideration in the conduct of affairs of the river authorities. . ... We therefore feel justified in our recommendation even if it involves the levying of precepts by bodies whose constitution is not based on comprehensive representation of authorities meeting the precept.[28]

The minority argued that the sub-committee recommendation for representation based upon user interest assumed that River Authorities would primarily operate as public utilities and that water use fees (abstraction charges) would soon become their principal source of financing.[29] The minority, on the other hand, assumed that a very large part of the River Authority funding would continue to come from land drainage assessments and precepts upon local authorities. Consequently, the minority held on to the conviction that local governments, as the agents responsible for levying taxes to meet precepts, should control River Authority management decisions. The essential elements of the minority's dissent are expressed in the following.

> The kind of authority, small and non-representative in character, which our colleagues favor would be an exclusive body akin to a commercial undertaking of one of the boards of a nationalised industry. River boards are not a trading organisation although they do need to take into account the economic results of their expenditure. They are a form of local authority providing a public service involving local knowl-

[28] *Ibid.*, p. 24.
[29] It should be noted, however, that user charges levied on water undertakers are passed on to retail users, usually through rates assessed against property valuations.

edge and understanding and neighbourly associations not only with similar authorities but with all sorts of people who live and work in their areas.[30]

They will have power . . . to precept upon County Councils [and] County Borough Councils and internal drainage boards [and] to levy general and special drainage charges. . . . To give these sweeping financial powers to a small non-representative body such as our colleagues recommend is without precedent. A precept made by a river board on a local authority is mandatory without any right of appeal. Thus the authorities called upon to pay would have no control . . . over the amount demanded and no voice as to how the money was spent. [Whereas] A right of appeal against water charges demanded by a river authority (with which we agree) will ensure that users of water are fairly treated. We think it quite wrong and entirely inconsistent, having provided justice for abstractors, to deny it, by non-representation, to local authorities . . . who will be required to find large sums to finance all the other activities of river authorities.[31]

The manner in which the River Authorities would be constituted was the only basic question on which the government in its White Paper challenged the sub-committee's recommendations. On this point the government essentially supported the minority.

On the issue on which the Sub-Committee was divided, the size of the river authorities, the Government agree with the majority of the Sub-Committee that there is much to be said for a small and compact body. On the other hand, much of the river authorities' income, particularly for their land drainage functions, will be raised by precepts upon local authorities, which therefore have a strong claim to be represented on the river authorities. The Government consider that a total membership of only 10–15 as recommended by the majority of the Sub-Committee, would be too small to provide sufficient representation of local authorities and at the same time to cover adequately the range of other interests which

[30] Sub-Committee on the Growing Demand for Water, *Final Report*, p. 33.
[31] *Ibid.*, p. 34.

will be affected by the river authorities' work and will contribute to their income.[32]

The White Paper then proceeded to outline the government's proposal: that the membership of the River Authorities consist partly of persons appointed by Ministers of central government because of their knowledge of particular aspects of the River Authorities' work or of interests affected by it, and that the members representing local authorities be in a bare majority. The government's proposal is reflected in the final legislation.

Representation in practice

The Water Resources Act of 1963 provides in some detail the manner in which the River Authorities are constituted.[33] A River Authority shall normally have not less than twenty-one and not more than thirty-one members. Thus the new river agencies are to maintain a smaller maximum size than the former River Boards, which, as mentioned above, often had forty members. The 1963 Act, however, permits the Minister to designate more than thirty-one members under special circumstances.

The law provides that: "Such number of members of a river authority as is sufficient (but not more than sufficient) to constitute a majority of the total membership of the authority . . . shall . . . be appointed by or on behalf of all the constituent councils." (The constituent councils are the councils of the administrative Counties or County Boroughs, any part of which is in the area of the River Authority.) Local government appointees, referred to as "local authority members," may be either members of constituent councils *or other persons*. The fact that local authority members need not actually also be members of constituent councils may result in their being less directly representative of and responsive to the taxpaying constituents than was perhaps presumed by the minority who argued for representation of those who are "called upon to pay."

[32] *Water Conservation, England and Wales*, Cmd. 1693 (HMSO, April 1962), p. 5.
[33] Essentially contained in Sections 6, 7, and 8 of the 1963 Act.

Perhaps the most significant feature regarding local member-ship is the fact that the law requires distribution of local repre-sentation to be based upon the taxpaying ability of constituent councils. "The number of local authority members . . . shall be determined by the Ministers having regard to the appropriate penny rate product[34] for the relevant area . . . for the relevant year." This provision seems clearly to establish taxpaying ca-pacity as a basis for local representation instead of special inter-ests, beneficiaries, or population criteria commonly used in the United States.

Non-local members of River Authorities are appointed by central government in the following manner:

1. By the Minister of Agriculture, Fisheries and Food, one or more "as being qualified in respect of" each of the following: (a) land drainage, (b) fisheries, and (c) agriculture.

2. By the Minister of Housing and Local Government, one or more "as being qualified in respect of" each of the following: (a) public water supply, and (b) industry (other than agriculture).

Members appointed by central government are required to be "qualified in respect of" the interests they are to represent, that is, "qualified as having had experience of, and shown capacity in, or otherwise as having special knowledge of, matters which relate to that subject as it affects the area of the river authority. In the appointment of local authority members of a river authority, the constituent councils are, so far as may be prac-ticable, to select persons appearing to them to have a practical knowledge of the matters to which the functions of the river authority shall relate." These two provisions suggest some am-bivalence as to whether members should be "representative" or "qualified" professionally in the fields from which they are designated.

An early proposal of the Minister[35] regarding the organization of the new River Authorities is instructive as to the possible

[34] Penny rate product is comparable to the use in the United States of "assessed valuation" as a measure of tax capacity of real property.

[35] Memorandum from W. R. Corrie, Ministry of Housing and Local Government, to the Clerk, Trent River Board, March 26, 1964.

implications of the representation provisions in the 1963 Act. Recommendations regarding membership were made for twenty-six Authorities (Isle of Wight River Authority was omitted). Of these, the median membership was twenty-five. Sixteen of the River Authorities had memberships of twenty-one, twenty-three, or twenty-five. The other ten Authorities were above this median group, two at the suggested maximum of thirty-one, and three had memberships proposed which would place them over the maximum suggested in the Act. In all cases, the local Authority members exceed the non-local members by one. Because membership recommended for the Trent River Authority is indicative of how the membership selection process might work in a specific case, an analysis of this proposal is presented in the Appendix which immediately follows.

SIGNIFICANCE OF MEMBERSHIP REQUIREMENTS: THE CASE OF THE TRENT RIVER AUTHORITY

This appendix analyzes the result of applying to the Trent River Authority the membership requirements of the Water Resources Act of 1963.[36]

The area under the Trent River Authority is the same as that under the Trent River Board. The River Board had forty-one members. For the River Authority the Minister proposed a membership of thirty-five, four over the suggested maximum. Local authority members were reduced from twenty-six to eighteen. One representative from agriculture, four from public water supply, three from industry, and one from navigation were to be added by designations from the appropriate ministers. Under this proposal, drainage and fisheries lost representation, numerically as well as relatively. (See Table 2.)

Of the eighteen local members proposed, ten were representative of counties and eight represented county boroughs (urban areas), for a county-urban ratio of 10:8. The county-urban area ratio in terms of total penny rate product is more nearly 10:11, while it is 10:7 in terms of population. This suggests that the proposed distribution between rural and urban jurisdictions more nearly reflects population ratios than it does the ratios of "penny rate product," which is prescribed as the basis of calculating representation. This observation is not based upon any definitive study. Perhaps its only significance at this time is to suggest the uncertainty of results of representation standards

[36] Most of the data and information unless otherwise indicated are taken from a memorandum from the Ministry of Housing and Local Government, "Water Resources Act, 1963, Establishment of River Authorities," March 26, 1964, which was sent to the Clerk of the Trent River Board, proposing membership for each River Authority. The proposed membership was subject to review, comment, and negotiation. The final data regarding the Trent Authority were not readily available when this report was completed.

TABLE 2. COMPARISON OF MEMBERSHIP FOR RIVER BOARD (1948) AND RIVER AUTHORITY (1963), TRENT RIVER BASIN

Interests Represented	Members of River Board[a] (1948)	Members of River Authority[b] (1963)
Local authorities	26	18
General	1	—
Land drainage	9	5
Fisheries	4	2
Agriculture	—	1
Public water supply	—	4
Industry (other than Agriculture)	—	3
Navigation (additional membership) Sec. 8 (3)	—	1
National coal board (additional membership) Sec. 8 (2)	1	1
Total	41	35

[a] Water Advisory Committee, Sub-Committee on the Growing Demand for Water, *Final Report* (HMSO, 1962), Appendix V.

[b] Memorandum, Ministry of Housing and Local Government, "Water Resources Act 1963, Establishment of River Authorities," March 26, 1964. Sent to Clerk of Trent River Board proposing membership.

that are applied to many component units rather than "at large."

The actual basis for local representation proposed for the Trent River Authority is shown in Table 3. The average penny rate product represented by each local member was £56,224. The average for representatives of counties was £53,400 and for county boroughs £59,755. Four county councils with an aggregate penny rate product of £17,430 were given no direct representation; two other counties shared a representative. Six of the fourteen county boroughs in the Trent River basin, representing £55,531 in penny rate product, were provided no direct representation. Under the above scheme, local members would represent penny rate product values ranging from £60,428 to £45,797, running from 7 per cent above to 18.5 per cent below the average of £56,224 per representative. The spread of £14,631 in the average values represented by each local member is only slightly over one per cent of the total penny rate product for the entire area. These data suggest that the proposed applica-

TABLE 3. PROPOSED REPRESENTATION FOR TRENT RIVER AUTHORITY,
BY PENNY RATE PRODUCT PER REPRESENTATIVE

Constituent Councils	Number of Representatives	Average Penny Rate Product per Representative
County Councils:		
Derbyshire, Leicestershire	3	£54,679
Lincolnshire, Yorkshire	1	47,945
Nottinghamshire	2	53,682
Staffordshire	3	50,474
Warwickshire	1	45,797
Average, County Councils		53,400
County Borough Councils:		
Birmingham, Solihull	4	53,479
Leicester	1	53,180
Nottingham	1	58,200
Smithwicke, Walsall, West Bromwich	1	60,428
Stroke-on-Trent	1	46,784
Average, County Borough Councils		59,755
Average for all constituent councils		56,224

tion of the 1963 Act to the Trent River Authority gives a fairly balanced representation on the basis of penny rate product.

This preliminary analysis detects a tendency for some over-representation of county councils as compared to urban areas under county borough councils; but this does not appear to be significant, particularly in view of the greater aggregate tax base and population under county councils. Normally it would be expected that the assessed value of taxable property and population would be greater in the county boroughs than in the general county areas. However, in the Trent Basin, this is not the case. The penny rate product for county councils is approximately 12 per cent more than the same assessment for county boroughs, and the population of the relevant counties is nearly 60 per cent greater than the aggregate population for the relevant county boroughs in the basin. Apparently this heavily industrialized Midlands area is essentially a continuous industrial complex, and in many instances counties are, in fact, highly industrial, heavily populated, and of course usually include larger areas. Since specific representation is prescribed by Ministerial Order, modifications to meet changing conditions can also be made by administrative action.

The significance of representation is ultimately measured by the way it affects voting power in the management board of the River Authority. The relative balance between county and county borough representatives may in reality be quite different from what it appears to be, depending on whether either or both groups vote as a block.

6

EVALUATION OF WATER MANAGEMENT INSTITUTIONS IN ENGLAND

In this concluding chapter the institutional arrangements for water management in England and Wales are appraised in terms of the six criteria developed in Chapter 2, and then are examined in a somewhat more general way.

Applying the Criteria

Ability to apply the total range of governmental techniques for influencing water resource use and development

This first, and basic, criterion relates to the extent and manner of legal authorization for each of the five techniques or stages of governmental involvement: (1) intelligence, (2) identification and planning, (3) regulation of water use, (4) development of water resources, and (5) regional distribution and disposal. With respect to this progression, the Water Resources Act of 1963 moved England a giant step forward.

Historically, the British had been largely dependent upon legal and administrative measures to regulate water use. The institutional system prior to the new law might in general be said to represent stage 3 of governmental involvement, as depicted in Figure 2. However, even the "intelligence" and "identification-planning" techniques had not been well developed.

Under the 1963 Act, the British system now possesses powers fully representative of stage 4 in the evolutionary model. All three techniques of the earlier stages—"intelligence," "identification-planning" and "regulation"—have been strengthened significantly. In addition, extensive and comprehensive authority has been added for resource development through the construction and operation of multiple-purpose projects and the promotion of co-ordinated river basin development.

A major factor in fulfilling this criterion is the creation of River Authorities. Ability to apply the range of techniques available in the law is greatly improved by the fact that the River Authorities are given primary responsibility for initiating authorized water management activities. Unification of water management responsibilities in the River Authorities should do much to promote a full and complementary use of the four basic techniques now provided in the English system.

Two deficiencies in the fulfillment of this criterion are significant. First, there appears to be inadequate authorization for organizing regional distribution and disposal systems. Although significant steps have been taken under powers given to the Minister of Housing and Local Government to consolidate water supply enterprises (private and public water undertakers), and to encourage extensions and combinations of sewerage systems, there is little evidence that the full potentials of regional economies of scale have been explored. Furthermore, since the administrative power for consolidating water distribution and disposal enterprises is lodged in the Ministry and thus isolated from the water management activities of River Authorities, it may be difficult, if not impossible, to design regional distribution and disposal systems hydrologically integrated with river management plans and actions. The Water Resources Act of 1963 gave little recognition to the question of regional distribution and disposal facilities. The possibility of River Authorities creating and managing subregional distributon and disposal districts with authority to purchase and consolidate existing systems and construct and operate facilities is foreclosed by failure of the Act to specify capital financing for this purpose.

A second deficiency is found in the authority to levy abstraction charges. If these charges are to provide incentives for efficient water use, and thus avoid the possibilities that they may unwittingly encourage inefficiencies, the present institutions for their administration in England seem weak on two counts:

1. Since the amount of the annual levy for abstraction charges is based upon annual revenues required to build and operate conservation works, it is unlikely that these charges can be so set as to moderate demand, either for water or for waste assimilation capacity. If this restriction were removed and annual charges more nearly reflected the value of water in use, abstraction charges could contribute to moderating demand in accordance with relative values of different kinds and amounts of abstraction.

2. The use of abstraction charges as an incentive for efficiency is further restrained by the manner in which the charges are passed on to individual users. Rates levied by water undertakers are subject to surveillance and adjustment by the Ministry of Housing and Local Government. There is little evidence that efficiency of water use has, or will be given, systematic consideration in setting "retail" water rates.[1] Moreover, even if it were, the Minister's authority is essentially independent and superior to the River Authorities and the Water Resources Board, thus close co-ordination in the administration of water charges as a tool of water management is difficult.

Ability to consider and adjust externalities
stemming from hydrologic interdependencies

The River Authorities in England essentially satisfy this criterion. Each Authority is given jurisdiction of an area delineated by combinations of river basins. Thus, externalities associated with surface water hydrology are brought to focus and can be internalized in the agency's decisions. The relationship between ground water systems and a River Authority area is not well known. Consequently the extent to which interdependencies between ground and surface water come within the

---

[1] This is not intended to imply that efficiency should necessarily be the only, or even the dominant, objective in setting water rates. In rate structures, various objectives of a financing scheme must be reconciled. However, the efficiency consequences must be considered, and it is not clear that present institutions will encourage that.

purview of a River Authority is indeterminable. However, water management powers of River Authorities apply to the ground as well as to the surface waters within their respective areas.

In addition, River Authorities have been given powers to co-ordinate non-Authority river projects within their respective areas—both by regulatory actions and by voluntary agreements and exchange payments. Four specific authorizations are important in this regard: (1) licensing of new or alterations of established non-Authority projects; (2) making agreements with water undertakers, local authorities and land occupiers, including financial exchanges, relating to construction, maintenance and operation of any works in the River Authority's area; (3) making agreements with navigation authorities, harbour authorities, and conservation authorities for exchange payments in accordance with damages incurred or benefits received by one from the projects of another; (4) approving exemptions or reductions of abstraction charges based upon spillover benefits resulting from the project of an abstractor. Each of these specific authorizations recognizes and provides a procedural remedy to potential externalities between the River Authority and other water development agents.

Both the scope and jurisdiction of the River Authorities and the procedural provisions between the Authority and non-Authority actions provide a favorable environment for adjusting externalities within each River Authority area.

Flexibility to adapt water management actions to different
circumstances of time and place with protection
against arbitrary and capricious action

Three considerations enter into an evaluation of flexibility: (1) the degree of administrative discretion provided in the law; (2) the extent that specific decisions are constrained by administrative goals, policies, and standards; and (3) the degree to which decisions can be reassessed and adjusted over time.

In England, the law gives the River Authority a wide area of freedom in exercising its water management powers. With few exceptions, an Authority is responsible for initiating within its area the major water management activities authorized. With regard to proposing a hydrometric scheme, minimum acceptable flows, a development plan, and a charging scheme, the law suggests time limitations within which the Authority should act. In these basic actions, approval by the Water Resources Board and/or the Minister is required. The law also provides systematic procedures for appeals, hearings, and reversals by higher authorities of most regional decisions. These latent authorities no doubt constrain the river agencies to some degree, but they are perhaps more important in the mind of the public as a protection against arbitrary and capricious actions.

Administratively promulgated goals, policies, and standards do not appear to threaten seriously the flexibility of decision making in the River Authorities. The British seem to have a highly developed practice of informal policy communications and there seems little inclination to the extensive use of formal policy directives. In pollution control, for example, they have resisted formally promulgating standards for the maintenance of water quality, even though the 1951 Rivers (Prevention of Pollution) Act authorized standards. In spite of reliance upon informal methods of policy influence, one potentiality has been slow in developing in England. This is the operation of independent research, investigative, and reporting organizations in the water policy field, such as occurs in universities and private research foundations in the United States.

Although the British seem to favor non-authoritarian policy devices, several specific types have emerged. For example, regional water supply plans, such as one recently completed for Southeast England, may become the instruments for shaping major policy structures within which the relevant River Authorities will be required to operate. It is within the power of the Water Resources Board to prepare and promulgate other regional and national plans, including those for transferring water from the jurisdiction of one River Authority to that of another.[2]

---

[2] One wonders whether plans prepared by the Board for large inter-Authority regional water schemes may not give undue weight to developing new sources of supply through construction of storages and diversions without adequate examination of the payoffs possible from less capital inten-

The Water Resources Board may, in fact, become the major source of policy constraints upon River Authorities with regard to many aspects of water management, even though the details by which it may effect its influence are not fully spelled out in the law. Certainly in its first three years the Board has been active and apparently effective in providing administrative and procedural guidance through the promulgation of a wide range of policy memoranda, often developed co-operatively with River Authority representatives. In addition to the Water Resources Board, the Authorities are required to operate under policies and procedures of the Minister of Housing and Local Government regarding general legal, financial, and administrative matters.

It should be noted that specific decisions may also be constrained by goals, policies, and standards established by an Authority itself. Thus, minimum acceptable flow determinations, water development plans, and charging schemes all represent a type of policy or standard within which subsequent decisions are presumably made. In each case, a general framework of conditions under which the River Authority will act is promulgated. For those who may be subject to specific River Authority decisions, published policies, plans, and standards provide a desirable degree of certainty and some assurance against arbitrary and capricious decisions. A balance between security for users and flexibility for the management agency is necessary.'

A major factor in the flexibility test is the extent to which goals, policies, and standards, once set, can be reviewed and revised. How institutions in England measure up to each aspect of flexibility must be deduced from the law. The statutes require that minimum acceptable flows and charging schemes be reviewed and revised as indicated within specified time periods. Consents to discharge waste water must be reviewed "from time

---

sive measures, such as instream reaeration, and more stringent regulation through licensing, zoning, and pricing. The extent to which development of new sources comes to receive undue weight may depend in large part upon whether central government subsidies for developmental works are expanded.

to time" and the River Authority may make "any reasonable variation" in the conditions attached to the discharge consent. Likewise, the authorization to license abstractions includes the power to revoke or vary the license upon satisfying prescribed procedures. This action can be initiated either by the Authority or the Minister. In these regards, flexibility appears to have been given consideration in formulating the laws.

It is not clear, however, what flexibility the River Authorities will have in the operation of their water development projects. The question here is whether the purpose or combinations of purposes for which a reservoir is designed might be changed as needs change. In the United States, for example, authorizations to build and finance reservoirs typically prescribe the purposes, and a question of legality may be raised if reservoir operations change the mix of outputs. The project construction and operation authority in England is comprehensive and without much qualification; however, financing is tied to purposes and later changes might be expected to raise questions, polictially, if not legally.

Ability to express and consider the range of values relevant to a water management decision

This criterion is concerned both with the accurate expression and consideration of economic demand for water services, and with the methods by which social-political "demands" are expressed and evaluated. Basically this criterion confronts the problem of articulating the demand systems and the governmental systems with supply considerations associated with the hydrologic system. (See Chapter 2.) Five institutional aspects are relevant to satisfying this criterion. These are: (1) the jurisdiction of the regional water management agency; (2) pricing and transfer payments; (3) representation on water management boards; (4) decision rules within such boards; (5) provisions for consultations, hearings, and appeals.

Jurisdiction. The areas over which River Authorities are given jurisdiction, whether by plan or by accident, tend to conform (or combination of Authority areas may conform) to designated planning and/or recognized economic regions.[3] This suggests that the River Authority, the areas of which are all delineated by hydrology, may actually subsume many of the relevant demand and governmental systems. Thus, the River Authorities may partially alleviate the disconformity problem discussed in Chapter 2.

Pricing and transfer payments. The system of abstraction charges in England provides a major linkage to the demand systems for withdrawal water. The extent to which this linkage will give the River Authorities a more accurate expression of the economic evaluation of water will depend in part upon: (1) the manner by which water undertakers are compensated by the ultimate user, and (2) the degree to which pricing will affect demand. In other areas, where pricing linkages are not feasible, intergovernmental transfer payments may serve to express public values associated with benefits rendered. This is particularly true if the receiving jurisdiction is free to negotiate the price. However, the English system of precepts among governmental jurisdictions is an assessment by the spending unit. The jurisdiction receiving the presumed benefit has no easy recourse to perceived overcharges. Apparently the only protection is through representation on the managing board of the jurisdiction which issues the precept. The question of local governments' representation on the River Authority is therefore of paramount importance.

Representation. The objective of representation is to give all parties at interest—beneficiaries and cost bearers—a voice in the decision process. In this regard, the River Authorities appear to be in a favorable position. Local members constitute a majority of the managing board and represent constituent county and borough councils in proportion to taxpaying capacity of areas within River Authority jurisdictions. Since water management has its major impact locally and regionally, and local taxes and charges constitute the major source of financing, representatives of local governments appear to be in a good

[3] At best, conformity is not complete, since River Authority boundaries are delineated by drainages, while planning regions are usually combinations of counties and county boroughs.

position to reflect differing values and negotiate reconciliations of conflict in demands.

The use of the local tax base as the criterion for distributing representation, does not seriously violate concepts of equal per capita representation when rural and urban populations are compared in the Trent River Authority.[4] (See appendix to Chapter 5.) Whether these results are indicative of the situation in other River Authority areas is not known.

The minority members, persons appointed by the appropriate Ministries on the basis of their special knowledge of specific water services (land drainage and flood control, fisheries, agriculture, industry, and water undertakers) presumably speak for the national point of view within these water use sectors.

Decision rules within River Authorities. Apparently little attention has been given to the rules by which River Authorities arrive at the decisions for which they are responsible. Does bare majority rule, or is some kind of consensus sought? The Water Resources Act, in an attached schedule relating to River Authority operations, merely says that an Authority may make such rules as it chooses with respect to its proceedings or to proceedings of its committees. This question was not explicitly investigated during these studies in England.

Consultations, hearings, and appeals. In general, the English law provides comprehensive and detailed procedures regarding consultations, hearings, and appeals. Requirements for formulating proposals for "minimum acceptable flows," "charging schemes," and "certain other orders and schemes" are contained in a separate schedule incorporated and enacted as a part of the Water Resources Act of 1963. In the case of MAF and charging schemes, for example, the Authorities are directed to notify and consult with an extensive list of specific interest organizations as a basis for developing proposals. Provisions for hearings and appeals are numerous, covering such important actions as MAF determinations, charging schemes, and levies thereunder, water rates levied by water undertakers, abstraction licensing,

---

[4] Neither tax base nor per capita representation necessarily reflects the intensity of values associated with interest groups concerned with water management.

reservoir licensing, and discharge consents. Appeals and hearings are usually thought of as devices to protect against arbitrary and capricious action, but they also serve to assure every party of interest a chance to be heard.

The English system appears to satisfy quite well all except one of the five institutional aspects of expressing and considering conflicting values. With respect to the fifth, prescribing decision rules for representative bodies, the law is silent.

Ability to finance water management consistent
with its objectives of efficiency

Two considerations have been identified as being relevant to this criterion: (1) unusual obstacles to raising capital and operating funds; and (2) the effect of the incidence of costs and benefits upon consideration of questions of efficiency. Both are difficult to assess definitively.

In general, the available methods of financing seem to present no unusual obstacles. Central government grants to subsidize water management activities are relatively more limited than in the United States, being principally limited to main river flood control (or drainage) projects and to the installation of approved hydrometric schemes. Although two general subventions —block grants and rate deficiency grants—may be used by local authorities for waste treatment and disposal facilities, it is unclear how large a factor they will play in the total field of water management.

Precepts are long established in British public finance, and apparently they successfully financed the more limited programs of the River Boards. However, it is yet to be demonstrated whether they will adequately finance those River Authority activities dependent upon them. Precepts are the basic method for tapping local tax resources, and tend to carry the residual financing load that is inappropriate or infeasible for grants or benefit assessments. For purposes authorized, there are apparently few restrictions on the amount of precepts that River Authorities may levy and which become an involuntary obligation of constituent counties and county boroughs to be met by additions to local tax rates. Most would agree that River

Authorities appear to have an unusual degree of freedom in financing from local tax sources.

Abstraction charges are new and are not required to be in full operation until 1969, four years after the effective date of the Water Resources Act. Since abstraction charges are expected to be the source of funds to support the new resource-conservation and development thrust, the four year delay in effecting charging schemes may appear to be an obstacle to financing. However, the law ameliorates this restriction in two ways. It permits the levy of abstraction charges on all new users (that is, those not qualifying for a license of right) even before a charging scheme is approved. The law also permits a River Authority, in the time between passage of the Act and approval of the Authority's charging scheme, to use precepts to raise capital and operating funds for conservation projects. In all respects, the abstraction charge appears to be a free and flexible instrument for financing activities that qualify under the rubric of water conservation. Annual levies can, within ceilings imposed by approved charging schemes, be set at levels necessary to meet any conservation obligation the River Authority chooses to assume. In terms of funding programs, the criterion for levying abstraction charges presents few obstacles; however, the basis for the annual levy has been criticized as restricting the achievement of efficiency objectives.

In terms of efficiency, two questions regarding financing methods are raised, one with regard to the impact of grants and the other with respect to precepts and abstraction charges. Subsidies in the form of grants from superior jurisdictions of government tend to discourage efficient decisions by the recipient. In general, the extent of disincentive is related to the per cent of total costs borne by the grant or, conversely, the proportion of local cost-sharing. In England, the amount of aid in any one drainage-flood control project is based upon judgment of the capacity of local authorities to finance the residual share that will be precepted upon them; thus these grants seem to have no efficiency motive. On the other hand, the fact that in practice over a ten-year period they average about 60 per cent of total costs indicates a higher local participation than has been typical in the United States. Another factor in amount of local participation which is not clear in the law is the extent to which grants from the central exchequer cover operation and maintenance costs. If, as inferred, grants are for capital improvements only, then local (River Authority) precept obligations are higher.

Both precepts and abstraction charges place a major incidence of cost upon the local taxpayer, either through general "rates" levied by local rating districts or through water "rates" or fees levied by water undertakers. Given the structure, the representation, and the strong local impact of the River Authority program, a high local incidence of cost would be expected to encourage efficiency in investment decisions by River Authorities. This is certainly true as compared to the situation if a greater and less discriminating use were made of subventions supported by the national taxpayer.

The fact that the law requires annual abstraction charges to be set on the basis of budget needs has been noted as a limit upon its effectiveness as an incentive to efficiency. The efficiency effect of abstraction charges will be further influenced by the manner in which charges are passed on down to the "retail" user. If, as is understood to be the case, water undertakers favor assessing their customers a water "rate," which like local authority rates is based upon property valuation, the individual, as the primary demand unit, has little incentive to adjust his demand to cost. Until greater emphasis is put upon metering and charging the ultimate user according to use, abstraction charges will fail to fulfill their potential as an incentive for efficient water use.

Although it is possible to identify these several lost opportunities to use the financing system to encourage greater efficiency, the fact remains that the financing system in England appears to be less conducive to inefficiencies than the usual practice in the United States where either explicit or hidden subsidies are a major factor.

Ability to conduct a continuing program of water management

The obvious care in the design and implementation of the Water Resources Act of 1963 is evidence of the intent to build water management into government in England. This characteristic of institutional arrangements stands in strong contrast to that in the United States, at state and particularly at federal levels. In the United States, emphasis is characteristically placed upon ad hoc studies and plans by governmental agencies, temporary committees, and/or commissions, with little or no attention to fundamental changes in the legal and administrative requirements relative to the basic techniques of water management. England's Water Resources Act and associated statutes, on the other hand, have effected significant changes in administrative authorizations relating to the five basic governmental control techniques and have placed these administrative activities in an institutional setting, which, to an impressive degree, satisfies the institutional criteria of this study.

Summary Evaluation

In addition to the specific evaluations above, several summary observations are in order.

First, the Water Resources Act of 1963 provides England and Wales with a remarkably comprehensive and apparently soundly conceived institutional system for water management. The Act is notable when compared to American efforts in that it succeeded in making legal, organizational, and financial adjustments and innovations at the same time. Any appraisal of the system in England must, of course, take into consideration the environment in which water management has come of age and in which it must now operate. Few would dispute that the new institutional arrangements (new legal powers and organizational structure within which to implement all relevant powers) provided by the 1963 Act represent a giant step forward in England's response to contemporary water problems. Not only is the present British system seen as an improvement over the past, but it has many attractive features when measured against the criteria of this study that might be emulated elsewhere.

In reviewing the system in England in terms of emerging concepts of the scope of water management, one is impressed with a second point. The new system still reflects the greater dependence on, and experience in, regulating water uses than in the development and management of the resource. All the conditions and procedures for regulatory activities, including details for notice, hearings, and appeals, are carefully prescribed in the statutes; while authorities regarding the new water development and conservation activities are quite summary. For example, the complexities of the water resource development function seem to be little recognized, particularly with respect to decision criteria and the importance of adequate analytical staff and planning procedures.

The inadequacy of legal powers to foster, or to construct and operate, regional waste collection and treatment systems is a third summary observation which has already been noted. Large unified waste disposal systems provide opportunities to use many of the newer technologies, permit the achievement of scale economies in providing disposal services, and facilitate integration of waste disposal into comprehensive schemes of water management. Regional waste collection and treatment systems are destined to become an increasingly important aspect of water management in heavily populated industrial centers. Yet it is in this area that British law seems most inadequate. At present, the River Authorities lack legal powers to finance construction and operation of regional waste collection and disposal schemes, or to purchase, consolidate, or otherwise coordinate existing systems. It may be predicted that the River Authorities will be handicapped in their aspirations to be regional water management agencies unless they obtain some control or influence over the design and operation of large unified waste management systems.

A fourth point that also deserves to be stressed is the restrictions in the law which inhibit the use of fees and charges to moderate demand and thus supplement administrative regulations. Again, the British law provides an unusually comprehensive system for assessing charges on withdrawn water. It appears to possess features which would permit it to operate as a regulator of demand (including demand for waste assimilation capacity), except that amounts of annual levies are restricted to those necessary to meet cost of conservation works. This statutory restriction will have to be removed if charging schemes are ever to be used effectively as a water management tool.

A fifth point concerns the organization of central government supervision. It should be noted that the Water Resources Board's role in supervising River Authorities seems to be well conceived, clearly prescribed, and well executed. Yet the Board's jurisdiction is ostensibly limited to only those new (conservation) functions with which the River Authorities are concerned. Several cabinet Ministers are concerned with other aspects of water resources. Their role in supervising River Authorities and the relationship between these Ministers and the Water Resources Board are not always clearly prescribed. A logical next step to improve the central-regional administrative relationships would seem to be to expand the scope of the Water Resources Board's concern to cover all aspects of water use and development. Notwithstanding this criticism, the apparent administrative linkages between the Water Resources Board and the River Authorities emerge as one of the more sophisticated central-field relationships in the area of water resource affairs.

A sixth observation emphasizes the fact that the Water Resources Act of 1963 has built into government a concept of integrated administration of the water resource activities. This stands in contrast to the "patchwork" that seems to characterize the approach to water resources in the United States at the state as well as at the federal level. In America, we have sought to provide integration by various devices for co-ordination. Major reliance is placed upon planning co-ordination, presumably on the assumption that if planning is co-ordinated, the implementation and operational aspects will fall into place as the spate of federal and state agencies pursue their normal activities. In England, policies, procedures, and institutional arrangements make planning one of the tools of a continuing responsibility for water management. England's planning can be dynamically related to changing needs from day to day, from year to year, and from decade to decade. Only time will tell to what extent the River Authorities and Water Resources Board make good the comprehensive planning authority that implicitly and explicitly interlocks their operations.

A final observation is that the present law in England is notable for the emphasis that is given to providing on a national scope regional integration of water resource actions through the River Authorities.[5] Whereas uniformity of policies and institutions as they serve regional and local needs has some attractions, it also may have some drawbacks in meeting the variety of needs for water and water resources development. A common pattern of policies and institutions may maximize certainty in the minds of individuals, and of private and public corporations. On the other hand, it possesses an inflexibility that may stifle experimentation and inventions in a field that needs to encourage innovations in both technology and institutions.

It is doubtful whether the United States could or should adopt a uniform, nationwide water policy and institutional pattern of the type now effective in England. The need for institutional flexibility may be greater in the United States, since the water resources themselves and the demands upon them have greater diversity among regions of the country. Perhaps the factor of greatest significance is the difference in size of river basins, with all the consequences of size upon the variety and complexity of physical problems and administrative solutions. These circumstances in the United States have seemed to make different approaches from region to region both more logical and realistic, while the relative similarity in hydrologic and development factors throughout England and Wales may justify common policy and administrative approaches. Thus, the institutional approach in England may have greater applicability within the separate American states than to the nation as a whole.

[5] It should be noted that Germany has long had permissive legislation, and France, even more recently than Great Britain, has enacted legislation providing for regional water management.